The DDT Myth

THE DDT

MYTH
TRIUMPH OF

THE AMATEURS

by RITA GRAY BEATTY

FOREWORD BY FRANCIS A. GUNTHER
Professor of Entomology and Chemist
University of California, Riverside

Published in Association with San Francisco Book Company

THE JOHN DAY COMPANY · NEW YORK
An Intext Publisher

Library of Congress Cataloging in Publication Data

Beatty, Rita Gray.
    The DDT myth.

Bibliography: p.
    1. DDT (Insecticide)   2. Pesticides—Environmental
aspects.   I. Title.
SB952.D2B4   301.31   72–12084
ISBN 0–381–98242–4
ISBN 0–381–90007–X (pbk.)

The John Day Company, 257 Park Avenue South, New York, N.Y. 10010

Published on the same day in Canada by Longman Canada Limited.

Printed in the United States of America

To my aunt MARY HOPP

Who taught me the real meaning of love
                    and
To my family

    My husband BEN

    Our children TOM, MARY, DANNY, JOHN

Contents

Foreword

A book of this type with courageous fact-seeking objectives is long overdue. From responsive conditioning to the public's taste for things dramatic, controversial, and emotional, the news media are generally inclined to emphasize the attention-getting "bad" rather than the prosaic "good," even in a field so complex, so dynamic, and so technical as modern pest control with chemical pesticides. When various individuals, groups, and subsequently organized campaigners decide to challenge the safety of a household item so long entrenched in our daily lives that it is well-known even to children, there can be only one result—a furor of allegations and charges followed by counter-charges equally emotion-laden so as to attract adequate attention.

If the attraction for entering this arena involves both the public health or welfare and the living environment on an imagination-staggering scale, the resultant contest can assume international proportions and consequences. As the initial foray gains depth and momentum through the skillful exploitation of dire emotionalism, whatever the cause, there will first be drawn into countless minor skirmishes as well as major well-publicized battles a horde of instant but usually well-meaning experts, followed by many deeply and justly concerned, puzzled individuals asking for the simple truth. This sequence is followed by responsive, knowledgeable, and concerned scientists who recognize the truth will be a complex rationalization based upon the only objective consideration of the issue possible— long-term risk versus long-term benefit, for short-term risks are excluded by law.

In the present instance modern pesticide chemicals have been vociferously attacked, for even the most casual observer is personally aware that these chemicals do indeed kill insects and weeds, the most visible annoyances to our comfort and home-gardening efforts, respectively. "These chemicals available everywhere in aerosol sprays, in supermarkets, and in nurseries were obviously designed to kill living organisms; ergo, they must be dangerous poisons to man," reasons the general public. Once having recognized this apparent truism, the intelligent public goes a step further and wonders about the effects of these "poisons" on cats, dogs, and fish ponds. The public is also made aware of the extensive agricultural uses of these chemicals and begins to wonder about the effects of their widespread use on birds and all other forms of wildlife, for we all have our cherished memories of having at least once seen a wild deer, a shy fox, a soaring eagle, or a trout leaping upstream.

When the anti-pesticide faction also claims irresponsible and unnecessary uses of chemicals in agriculture and forest management, and then caps the allegations with the specter of cancer, there can be no alternative but thorough and systematic fact-finding with evaluation by competent persons.

The deliberate use of pesticide chemicals dates back to antiquity when man first recognized that smoke would repel some pestiferous flying insects, as well as preserve meat against spoilage organisms. In the *Odyssey,* Homer wrote about 800 B.C. of "pest-averting sulfur." By 1940, we had about two dozen pesticide chemicals, each effective against only a very few pests, and usually for very short periods even at high dosages. For example, lead arsenate was required for spraying apple trees for codling moth control at a rate of twenty pounds per acre, with four to twelve applications per season. The only general purpose insecticides at that time were the extremely toxic hydrogen cyanide (as a fumigant, with many deaths each year among applicators) and nicotine (less than one g., which is about 20 drops, has been fatal for man; there were many deaths each year among applicators).

These pesticides were regulated as to standards of purity and usefulness by the Federal Insecticides Act of 1910. Between 1942 and 1945 the novel "wide-spectrum" (almost all-purpose) low-dosage, and long-effective insecticide DDT was introduced with spectacular success into pest control. It was followed by BHC (and its active component lindane), chlordane, toxaphene, heptachlor, aldrin, and dieldrin by 1950. These are the persistent chlorinated hydrocarbon (organochlorine) or "hard" insecticides. Beginning about 1948, the first "wide-spectrum" organophosphorus or "soft" (less persistent) insecticides were broadly introduced. These included parathion, malathion, schradan, TEPP, demeton, and so on, followed by many others, some with very narrow "spectrums" of activity. By 1960 we had about 400 major pesticide chemicals (insecticides, miticides, herbicides, fungicides, nematicides, rodenticides, etc.), most of which were again effective against only certain pests, but which were also generally used at low dosages with few applications per season; these materials were then regulated by the Federal Insecticide, Fungicide, and Rodenticide Act of 1947, which required registration, proof of efficacy, and absence of discernible public health hazards when used as directed. In 1954 the Federal Food, Drug, and Cosmetic Act required that petitions for the sale and use of pesticide chemicals must include full data on their toxicity and other aspects relating to public health. The 1947 and 1954 Acts are applied jointly, so that an extremely hazardous chemical will be denied registration and tolerance assignment (a permitted safe level in foods, well below the maximum acceptable daily intake for human beings). Subsequent amendments and regulations have tightened these controls even more, prohibiting the use in food production of any chemical suspected of or known to cause cancer and requiring extensive data upon the effects of any pesticide chemical upon wildlife and of the migration of that chemical into the environment away from the site of its application.

Because all pesticide chemicals are considered "poisons" by the public, and because "poisons" is a fearsome word, numerous popular writers and speakers have questioned the wisdom and value of these chemicals. The first major journalistic attack against pesticides was William Longgood's book *The Poisons in Your Food* in 1960. In 1962 the readable and persuasive *Silent Spring* by the eminent marine biologist Rachel Carson sounded the appealing call-to-arms for those individuals genuinely concerned about the necessity for these "poisons." Miss Carson focused commendable attention on possible hazards to be associated with the widespread use in modern society of *any* chemical more toxic than salt or sugar, and she assailed pesticide chemicals with both fact and fancy. Because of its spectacular performance and its extensive and long use for so many purposes, she was particularly vociferous against DDT. As with most authors, she was also inclined to favor arguments that advanced her thesis of doom and to minimize those that did not.

The next few years brought some small-scale skirmishes between the pros and the cons. Most adequately informed scientists were certain that reason would prevail because the benefits from the controlled uses of pesticide chemicals so obviously outweighed the minor risks that were carefully and elaborately minimized by both state and federal legislation, regulation, and surveillance. This naïveté has been most costly, with the eventual incredulous realization by those scientists genuinely concerned with both the public health and the environment that fancy can prevail over dull fact when presented by a skilled elocutionist artful in offensive rather than defensive tactics. Most anti-pesticide advocates forget that the informed scientists promoting the regulated use of pesticide chemicals when necessary to protect health or crop are as environment and health conscious as anyone else: like the pilot of a commercial aircraft they too feel that they have a vested interest in a safe flight.

The Wisconsin DDT hearing in 1968 rallied forces for the all-out defense of the environment and the public health, and the battle was joined on both sides by the informed, the misinformed, and the partially informed. DDT was the target of the day. DDT was known to everyone as a household word. If the anti-pesticide crusaders had chosen as a target azinphosmethyl or amidithion or carbophenothion or trichlorfon or demeton, *ad infinitum* through 400 pesticide chemicals, the public would have said "Ho hum," for who but specialists ever hears of these currently important pesticides? DDT, however, evoked an instant response by everyone: "DDT? I have some in the house in an aerosol can! I use it once a week! I didn't know it was killing us!"

Rita Gray Beatty, the author of the present volume, is a professional journalist, not a scientist. A mother herself, with an inquiring and skeptical mind, she too began to wonder about three years ago about DDT in mother's milk, in cow's milk, in the environment, in wildlife, and in rain and arctic snow, for she too read newspapers and magazines and watched newscasts and debates on television. Recognizing inconsistencies in some of the claims and counter-claims in the DDT story, she was intrigued to the point of an intensive investigation to separate fact from fancy. This book is the result of more than two years of really extensive efforts to resolve the two DDT myths: "DDT is the greatest blessing ever to come out of a chemical laboratory" versus "DDT will yet be the bane of mankind." Hundreds of personal contacts, letters, telephone interviews, and literature comparisons later she decided as a concerned citizen that

DDT was indeed being viciously maligned by some, generously abetted by countless camp followers, well-intentioned but inadequately or improperly informed.

She concluded that the present story must be told in a balanced way to other concerned citizens and in a language comprehensible to non-scientists: during this amassing of information she had also discovered that all scientists talk and write in technical jargon which requires interpretation for adequate comprehension by the uninitiated. I and many others have had the pleasure of helping her learn the jargon and intricacies of modern pest control. This book, however, is Mrs. Beatty's own synthesis. It represents her opinions and conclusions, a jigsaw puzzle laboriously but conscientiously and impartially assembled to present the many-faceted but actual picture of the DDT myth according to available documented information. This book compares fact with fancy; it points out both strengths and weaknesses in modern pest-control practices; it chides the camp follower for impetuosity; it chides the scientist for taciturnity; it chides the agricultural chemicals industry for failure to recognize adequately that the public has a right to be informed on matters affecting their continued existence on a hospitable planet; it chides some governmental agencies for elements of duplicity and procrastination; and with one notable exception it rebukes the news media for a strong tendency to accentuate the "bad" and to minimize the "good." This book represents a plea for sanity, prudence, and proper perspective on the part of all of us—the public *in toto* must be protected from undue hazard commensurate with an acceptable standard of living plus the greatest possible health benefits, and this standard of living must incorporate all aspects of a hospitable environment delicately imbalanced in favor of man.

I have been professionally involved with DDT in our food supply and other segments of our total environment since 1942–1943, when the U.S. Department of Agriculture shared with me the first sample to arrive in the United States from neutral

Switzerland. My first technical publications on DDT were in 1945 after this exciting new pest-control chemical was released by the Defense Department from its "top secret" classification. In 1946 I publicly raised the question of its fate in the environment at an annual meeting of the Pacific Slopes Branch of the then American Association of Economic Entomologists (now the Entomological Society of America). Both as a pesticide residue chemist and as a concerned citizen, I, therefore, recommend this book to everyone curious to learn how easily we can be beguiled and how laboriously and ponderously we then unravel Shakespeare's "tangled web we weave."

Dr. S. A. Forbes, entomologist, 58 years ago wrote the following:*

"The struggle between man and insects began long before the dawn of civilization, has continued without cessation to the present time, and will continue, no doubt, as long as the human race endures. . . . We commonly think of ourselves as the lords and conquerors of nature, but insects had thoroughly mastered the world and taken full possession of it long before man began the attempt. They had, consequently, all the advantage of a possession of the field when the contest began, and they have disputed every step of our invasion of their original domain so persistently and so successfully that we can even yet scarcely flatter ourselves that we have gained any very important advantage over them. . . . If they want our crops they still help themselves to them. If they wish the blood of our domestic animals, they pump it out of the veins of our cattle and our horses at their leisure and under our very eyes. If they choose to take up their abode with us we cannot wholly keep them out of the houses we live in. We cannot even protect our very persons from their annoying and pestiferous attacks, and since the world began, we have never yet exterminated—we probably never shall exterminate—so much as a single insect species.

*From *Destructive and Useful Insects,* by C. L. Metcalf, W. P. Flint, and R. L. Metcalf, McGraw-Hill Book Company, New York, 1951.

"They have, in fact, inflicted upon us for ages the most serious evils without our even knowing it."

This picture has not really changed, has it?

January 10, 1973

FRANCIS A. GUNTHER
Professor of Entomology and Chemist
University of California
Riverside, California

Preface

This book has been written completely at my own expense. I have received no financial support from any company, corporation, institution, organization, or individual. I am deeply indebted to the many scientists who have assisted me by supplying background information and helping me to unravel the complexities of the scientific jargon. None of these men and women have contributed financially to the researching of the book, nor are they responsible for its contents except in those cases where they are directly quoted. I have no vested interest in any company or organization, nor am I employed by or affiliated with any organization whatsoever. I have in my possession copies of all sources cited in this book, with the exception of personal interviews and telephone interviews.

This book has been written without malice toward any group or individual. It is the painstaking result of piecing together information which generally is unavailable to most people.

My husband and I have four children who have frequently asked the same question regarding a variety of situations, "Why doesn't somebody do something about it?" Our answer has always been, "*You* are somebody. Why don't *you* do something about it?" Many Congressmen, city officials, and toy manufacturers have had to decipher the questions and complaints of our children from the time they were old enough to write. This book is the result of practicing what I have preached.

Mountain View, California
January 3, 1973   RITA GRAY BEATTY

Acknowledgments

Acountless number of men and women—those associated with government agencies and departments, private industries, universities, hospitals, and national and world organizations, as well as numerous private specialists in many fields—have provided background material, advice, counsel, and moral support in the preparation of this book. Although it would be impossible to list by name the hundreds of individuals who have given so generously of their time to assist me, each contribution has added something of value to the book and is deeply appreciated. Naturally, the final responsibility for the accuracy and validity of the text is mine.

In particular, I would like to thank the following people, whose expertise was invaluable to the completion of this book.

Essential data were provided by these members of various Federal, State, and County Government Agencies: Commander R.O. Baker, United States Navy; Honorato S. Echavez, Industrial Hygienic Engineer, Los Angeles County Health Department; Dr. Owen A. Kearns, M.D., Health Officer, Kern County Health Department, Bakersfield, Calif.; Dr. Robert M. Linn, Chief Scientist, National Park Service, United States Department of the Interior; Dr. Alice Ottoboni, Toxicologist, Department of Health, State of California; Donald Robinson, Superintendent, Channel Islands National Monument, United States Department of the Interior, Oxnard, Calif.; Congressman Jamie L. Whitten, Chairman, Subcommittee on Agriculture, House Committee on Environmental and Consumer Protection.

Valuable information on various studies conducted by private industry was provided by the following persons: Hal Barber, Monsanto Company, Los Angeles; Dr. R. O. Clinton, Chemist, Sterling Drug, Inc., New York; Dr. Cortez M. Enloe, Jr., M.D., editor, *Nutrition Today*; Dr. Richard L. Hall, Chemist, McCormick-Schilling Co., Cockeysville, Md.; E. V. John, John C. Livengood, and W. B. Papageorge, Monsanto Company, Mo.; Jack Pressley, IMC Food Products, Ill.

Research of vital importance has been conducted at various universities and hospitals, and these doctors, professors, and professional people were particularly generous with their time and knowledge: Dr. Ellsworth C. Alvord, Jr., M.D., Professor of Pathology, University of Washington, School of Medicine, Seattle, Wash.; Dr. L. D. Anderson, Professor of Entomology, University of California, Riverside; Dr. Jay M. Arena, M.D., Professor of Pediatrics, Duke University Medical Center, Durham, N.C.; Thomas G. Beatty, Associate Editor, *California Engineer,* University of California, Berkeley; Dr. Bruce Beckwith, M.D., Neuropathologist, Children's Orthopedic Hospital, Seattle, Wash.; John D. Furrer, Agronomist, USDA Extension Service, University of Nebraska, Lincoln; Dr. Louis Gluck, M.D., Professor of Pediatrics and Obstetrics, University of California, San Diego; Dr. Wayland J. Hayes, Jr., M.D., Professor of Biochemistry, School of Medicine, Vanderbilt University, Nashville, Tenn.; Dr. Edward R. Laws, Jr., M.D., Resident-in-Chief, Neurological Surgery, The Johns Hopkins Hospital, Baltimore, Md.; Richard Main, Biologist, California State University, San José, San Jose, Calif.; Dr. H. T. Reynolds, Professor of Entomology, University of California, Riverside;

Dr. Robert M. Schuman, M.D., Neuropathologist, University of Washington, School of Medicine, Seattle, Wash.; Francis W. Smith, Biologist, Southern Oregon College, Ashland, Ore.; Ms. Claire Still, Librarian, Stanford Medical Library, Stanford, Calif.; Dr. Philip Sunshine, M.D., Director of Newborn Care, Stanford University Medical Center, Stanford, Calif.; Dr. W. E. Westlake, Research Chemist, University of California, Riverside.

Invaluable assistance has been lent by these members of world scientific organizations: Dr. Norman E. Borlaug, Agronomist, Centro Internacional de Mejoramiento de Maiz y Trigo, Mexico City; Dr. Guzmán García Martin, Chief, Malaria Eradication Department, Pan American Health Organization, Washington, D.C.; Dr. N. G. Gratz, Entomologist, World Health Organization, Geneva, Switzerland; Dr. A. G. Ponomarenko, U.S.S.R. Academy of Science, Paleontological Institute, Moscow, Russia.

In addition, I am grateful for the expertise of these private individuals: Frank L. Beebe, Specialist-Peregrine Falcons, Saanichton, British Columbia, Canada; H. R. Conklin, Horticulturist, Cashmere, Wash.; Dr. Thomas P. Fuller, M.D., Mountain View, Calif.; Ms. Setsuko Hirano, Mountain View, Calif.; Dr. G. E. Quinby, M.D., Wenatchee, Wash.; Dr. Richard T. Rappolt, Sr., M.D., Clinical Toxicologist, San Francisco, Calif.; Dr. Joseph Still, M.D., West Covina, Calif.

Finally, there are five men without whose help and continued support this book could never have been written. My special thanks to them:

Dr. J. Blair Bailey, Pesticide Safety Specialist, Division of Agricultural Sciences, State of California, Member of the President's Cabinet Committee on Occupational Exposure to Pesticides;

Benjamin W. Beatty, Biologist, Physiologist, Los Altos High

School, Los Altos, Calif.; he is virtually the co-author of this book;

Dr. Walter Ebeling, Entomologist, Department of Agricultural Sciences, University of California, Los Angeles, Calif.;

Dr. J. Gordon Edwards, Professor of Entomology, California State University, San José, San Jose, Calif.;

Dr. Francis A. Gunther, Professor of Entomology and Chemist, University of California, Riverside.

The DDT Myth

A Game
of
Dominoes

1

An immense accumulation of inaccurate information and emotionalism about pesticides, especially DDT, has placed the United States, and indeed the world, in an untenable position. Powerful and articulate groups of misguided environmentalists have sought, and doubtless will continue to seek, bans on the use, production and distribution of DDT. These groups attributed a wide range of effects to its use—from thin eggshells to the pollution of mother's milk, from a lack of female fertility to unlinking the food chain. DDT is in fact responsible for none of these things, as will be shown.

There is not enough DDT in existence now to take care of the critical needs of the world's population. A ban or curtailment on its production, interstate shipment or essential uses would have far-reaching and catastrophic effects not only in the United States but also throughout the world.

In 1948, Paul Müller was awarded the Nobel Prize for his discovery of the lifesaving effect of DDT. Dr. Norman E. Borlaug was awarded the Nobel Peace Prize in 1970 for his efforts to ease world hunger through the development of high-yield wheat strains. In October, 1971, Dr. Borlaug flew to Washington, D.C., at his own expense from his home in Mexico City, to testify before the Environmental Protection Agency Hearings in favor of the continued use of DDT. The following is an excerpt from Dr. Borlaug's statement:

Environmentalists today seek a very simple solution to very complex problems. The pollution of the environment is the result of every human activity as well as the whims of nature. It is a tragic error to believe that agricultural chemicals are a prime factor in the deterioration of our environment.

The indiscriminate cancellation, suspension, or outright banning of such pesticides as DDT, is a game of dominoes we will live to regret.

DDT, because it is a name popularly known to most segments of the public, has been the first target. Once this is accomplished, the so-called ecologists will work on hydrocarbons, then organo-phosphates, carbamates, weed-killers, and, perhaps, even fertilizers will come under the assault of their barrage of misinformation.

If this happens—and I predict it will, if most DDT uses are cancelled—I have wasted my life's work. I have dedicated myself to finding better methods of feeding the world's starving population. Without DDT and other important agricultural chemicals, our goals are simply unattainable—and starvation and world chaos will result.

An article in the *Wall Street Journal,* February 16, 1970, read, "A spokesman for the United Nations Food and Agriculture Association [Organization] has stated that if the proportion of crop lands in India, now protected primarily by DDT, were to be increased from 10 percent to 20 percent, India could grow an additional 1,400,000 tons of rice, 100,000 tons of peanuts, 65,000 tons of sorghum, 250,000 tons of sugar, 46,000 tons of corn, and 200,000 tons of potatoes each year. This spokesman said, 'It is hard to talk about pesticide contamination in food to a man who is starving to death.' "

There is much more at stake. DDT has frequently been called a health hazard and yet it has saved as many if not more lives than penicillin. The World Health Organization (WHO) has steadfastly maintained that DDT is the safest, cheapest, and most effective agent in the eradication of insect-borne diseases. Dr. James Wright, Chief of Vector Biology and Control for WHO, has stated: "DDT has been the main agent in eradicating malaria in countries whose population totals 550 million people, of having saved five million lives and prevented 100 million illnesses in the first eight years of its use, of having recently reduced the annual malaria death rate in India from 750,000 down to 1,500 and of having served at least two billion people in the world without causing the loss of a single life by poisoning from DDT. It is so safe that no ill effects have been observed among the spraymen or among the inhabitants of the sprayed areas."

A few years ago the average American did not have the word *ecology* in his vocabulary. Today millions of people have become self-appointed ecologists without the years of academic study and experience required. Widespread interest in ecology is good, of course; but scientific decisions should be left to those men and women with the background and knowledge needed to make such decisions based upon scientific evidence rather than half-truths and innuendos. Ecology today is like a giant swimming pool. Everyone wants to get into the water, but not everyone can swim.

Ecology is the study of the way plants and animals (including man) affect each other and interact with their inanimate surroundings. In so complex an area of study with its numerous subdivisions, it is important that we rely upon informed sources rather than amateurs. Arthur Godfrey is a case in point. He has persistently denounced the use of DDT and attributed countless hazards to its use. An example is this pronouncement from *The Arthur Godfrey Environmental Reader:* "I came across a little horror a while ago that has stuck with me. Know why the sea turtles are dying out? One big reason is the plastic bags messy human beings are throwing overboard. The turtles think the bags are jellyfish and swallow them; then they choke to death.

"This is pretty dumb of the turtles, maybe—but aren't we doing the same thing, eagerly swallowing each and every technological 'advance,' and then choking on it? DDT, that kills dangerous insects . . . and birds . . . and animals . . . and people. You name, we try it—try now, pay later—cars, nuclear energy, detergents, plastics."

In a television commercial a man and a dog were shown walking along a clean sandy beach. The man, claiming to be a medical doctor, told us that because of his *concern for our environment* he and a group of doctors had developed a new washday product called ECOLO-G. He said, "If you care about your environment, stop polluting our water with regular detergents, use ECOLO-G." This commercial appeared for several weeks and then the news media carried a notice to housewives about the product. The Food and Drug Administration had found that it contains sodium metasilicate, which causes a red rash and is toxic. The product is still on the market, but now carries the following warning:

DANGER – MAY CAUSE BURNS – HARMFUL
IF SWALLOWED – EYE IRRITANT
KEEP OUT OF REACH OF CHILDREN

The time has come to place things in proper perspective. Emotionalism against DDT continues to mount. For the most part, a one-sided picture has been presented. It is vitally important to people of all nations that we all seek the truth.

Nothing is poison, yet everything is poison. It is all a matter of degree. Even distilled water can kill if too much is drunk too rapidly by an individual who has been lost in the desert for an extended period of time. If a bird were to be fed an excessive amount of regular table salt mixed with its food, the bird would die. At the opposite end of the spectrum, very small amounts of "highly poisonous" arsenic compounds and the common explosive nitroglycerine are regularly and purposely taken by many human beings with beneficial results. It would be inaccurate to say that DDT is not a poison. It could be if it were improperly used. The same can be said of salt, water, aspirin, nitroglycerine, *ad infinitum*.

Dr. Marshall Laird, chairman of the biology department, University of Newfoundland, told the United States House Congressional Committee on Agriculture in March 1971 that medical entomologists do not have a single insecticide ready for use that is capable of providing levels of control equivalent to DDT.

Joseph W. Still, M.D., has studied the overall effects of DDT since it was first introduced by our armed forces in Italy in 1942–1943. In a letter to William D. Ruckelshaus, the head of the Environmental Protection Agency, January 18, 1971, Dr. Still wrote, "The annual Audubon Society counts of all the major birds including robins and songbirds have steadily risen since DDT was introduced into the U.S. Similarly all the major game animals have shown a parallel increase. I believe this is due to the fact that DDT kills the insect pests that carry diseases to birds and animals. So instead of being an enemy of birds and animals, it is actually their benefactor."

The president of the National Plant Board, Dr. Harold Porter, sums up the general position of a majority of the scientific community: "We view with increasing alarm the growing reaction against pesticides and the increasing restrictions on their use. We now have several pests which are truly serious threats to our forest and food crops, for which we do not have really effective controls other than some of the pesticides currently under fire." Dr. Porter concluded that many of the materials we are forced to use in place of DDT are so toxic that farmers and spraymen refuse to use them.

On March 18, 1973, Dr. Francis A. Gunther, Professor of Entomology, University of California, Riverside, said, "The federal government now has a massive, multi-million dollar, nation-wide, crash program in operation to train farmers and farm workers how to handle and how to apply these DDT substitutes without killing themselves and their neighbors. The agriculture experiment stations and the United States Department of Agriculture extension services are hard at work across the nation to implement this extremely extensive program."

This is meritorious; nevertheless, vast problems remain. Will farmers and farm workers be aware of this program? Will they take advantage of it—indeed, will they be *able* to take advantage of it? It is obviously impossible for the scientists to reach every farmer and farm worker. It is equally impossible for every farmer and farm worker (particularly the migrant farm workers) to reach the scientists.

Since the well-being of mankind, according to Dr. Borlaug and countless other eminent scientists, is at stake in the controversy over DDT, should we not insist that all judgments on DDT be based upon valid, documented scientific evidence rather than upon the emotional, frenetic, and too often faddish opinions of amateurs?

The War on Disease

2 There is enormous interest in the United States at the present time in everything "natural." We are asked in the name of protecting the environment to rid ourselves of all "unnatural" aids to mankind. To do so would be to sow for harvests that millions of people would not live to reap. Although there are in our culture many artificial elements that require constant review, we must not abandon the basic resources for survival that man has developed for coping with nature, which is not benevolent. Nature is lovely, soul-lifting, and deadly.

DDT has saved millions of lives in battles with nature. It can save millons more if it is not eliminated from man's arsenal against disease and pestilence.

Dr. Thomas H. Jukes, Professor of Medical Physics, University of California, Berkeley, has said it all in "A Town in Harmony," an article about a fictional American town of an earlier century.

The harmony of all life in this idyllic town followed a biological balance in Nature, a balance which man had not yet learned to disturb by drastic intervention on his own behalf.

As the sun went down, the buzzing of mosquitoes could be heard in town; the malaria parasites in their salivary glands were about to continue their life cycle in the red blood cells of human victims.

The last slanting rays of the sun lingered on the small headstones in the town graveyards. Here slept the children who had perished from diphtheria, scarlet fever, and whooping cough. Beside them lay the bodies of those who had lived in harmony with proliferant typhoid germs. These bacteria, uninjured by chlorine or antibiotics, teemed in the limpid stream that ran at the edge of town. It flowed through meadows where grazed cows beneath whose hairy flanks swarmed trillions of tuberculosis organisms, waiting for their milky ride that would take them to the lungs of the townspeople.

Life for these folks was a struggle with unrelenting nature. Some of them, including the Carney family, had fled from Ireland at the time of the potato famine, when a fungus disease had turned the food supply into a stinking black slime, so that many people had perished from starvation and from diseases resulting from starvation.

Famine had not laid its hand on the New World, but farmers knew what it meant to see a good field of wheat flattened by stem rust and the Hessian fly, a scabby and wormy apple crop lying on the ground; to see rows of young corn destroyed by cutworms and wireworms, pigs dead from hog cholera, and Canada thistles choking out the oats as the white, fluffy seeds sailed on the breeze to the next pasture.

It had been a warm afternoon, and a hush had settled on the grocery store. Faint sounds could be heard; a friendly rat gnawing in the cellar; the rustle of the weevils in the cracker barrel; the high-pitched buzz of flies that were struggling in the sticky festoons hanging from the ceiling, and the stealthy patter of the cockroaches that darted across the floor.

Yes, life was in harmony with its surroundings. The women who, a century later, might have been writers of science fiction horror stories, were too busy with their housework to read humor magazines. They were squashing black beetles; beating the clothes moths out of the winter woolens; scraping the mold from the fatback pork; and wondering if they could afford the luxury of a chicken for their Sunday dinner.

Man's greatest biological enemies have always been and continue to be disease and hunger. Of the countless diseases that have plagued man, the insect-borne diseases remain today the most difficult to control.

MALARIA   Malaria is still the number one killer of mankind. It affects one-sixth of the human race today, killing one person every twelve seconds. Malaria is a disease which has plagued mankind for centuries. One of its many nasty side effects is that it has a tendency to return for a second bout with those victims who did not die in the first round. Malaria causes chills, extremely high fever, extensive liver and kidney damage, repeated attacks, and death.

The female anophelene mosquitoes, which carry malaria, are the bug world's equivalent to vampires. They too suck blood for food. What happens is a chain reaction: An infected mosquito bites a healthy man, introducing swarms of malaria parasites into his bloodstream. After a short incubation period the man develops malaria. Healthy mosquitoes then bite this now malarious man and having sucked some of his blood, rest on the walls inside the house for two or three days to digest their meal. During this period of time these previously healthy mosquitoes become carriers of malaria, and when their hunger pangs strike again, they fly off to bite other people. Thus they continue the destructive chain. Once malaria breaks out in an area it can spread like wildfire.

The best known way to break this chain is to spray the walls and ceilings of the houses with DDT. This procedure kills the resting mosquitoes and prevents them from adding new links to the chain.

By 1968, the United States had spent more than half a billion dollars to help control malaria throughout the world. As a result, more than 960 million people who a few years ago were in constant danger from malaria are now free of the disease because DDT can effectively kill the anophelene mosquitoes that carry malaria without causing harm to the people. Another 288 million people live in areas where this disease is being so vigorously attacked by DDT that malaria will be eradicated in time.

Dr. Guzmán García Martin, Chief of the Malaria Eradication Department for the Pan-American Branch of the World Health Organization, has stated that the withdrawal of DDT, the main weapon in the world-wide antimalaria battle, would mean the interruption of most malaria prevention progress throughout the world. He warns that the withdrawal of DDT would result in disaster because large-scale epidemics of malaria would follow and millions of people would die before enough DDT could be manufactured and distributed to bring this disease under control again.

The disaster in Ceylon in the 1960s is an example of how quickly and extensively malaria can return to an area if the use of DDT spray is not continued until the disease is completely controlled. Ceylon, an island of modest size off the southern coast of India, has a population of about twelve million. Of these twelve million inhabitants, about nine million are farm workers. Ceylon has 25,332 square miles of land, only one-fourth of which is habitable; thus there are approximately two thousand persons per square mile.

Following a nation-wide eradication program with DDT in the 1950s and early 1960s, the number of malaria cases in Ceylon reached lows of thirty-one in 1962 and seventeen in 1963. The program was eliminated in 1964.

Malaria increased annually thereafter, numbering 150 cases in 1964, 308 in 1965, 499 in 1966, and 3,466 in 1967. In 1968 total havoc broke loose: 16,493 cases were reported in January, and 42,161 in February. No DDT supplies were available to reinstate house spraying programs on the scale needed; consequently more than one million cases of malaria occurred in 1968. Remember, Ceylon had a *total* population of only twelve million. At this point the government of Ceylon wisely decided to return to the use of DDT, but, not so wisely, the order was placed in an Eastern European country which did not deliver. A year later Ceylon placed an order with the United States for nearly ten million pounds of DDT. The order was marked "Urgent." Unfortunately, a dock strike held up the shipment for months, but obviously it did not hold up the mosquitoes in Ceylon. They worked overtime.

Before the World Health Organization (WHO) began the antimalaria campaign in India, an estimated one million people died of malaria each year. An additional 75 million persons contracted malaria and were left too sick and weak to work. In an underdeveloped country, if the people do not work, they do not eat. If malaria did not kill its victims, starvation did.

As a result of the malaria epidemics, India was losing at least two million work years annually. Thanks to DDT, millions of Indians are healthier today, and thousands of square miles in India (as well as in Africa) that once lay fallow are under productive cultivation.

Over a thousand other chemicals have been tried for use in malaria control by WHO, the Pan-American Health Organization (PAHO), and the National Communicable Disease Center (NCDC). Only three pesticides other than DDT have been found useful, and in only one percent of the total program. They are malathion (an organophosphate which is readily absorbed through the skin), arprocarb (a carbamate, also known as propoxur or Baygon, which is costly and more toxic to man than DDT) and fenitrothion (an organophosphate, also known as Sumithion, which is absorbed readily through the skin and is expensive). Another problem with these three chemicals is that they must be applied frequently, and only by skilled, well-protected workers.

WHO recommends the continued use of DDT. Since about one percent of the anophelene mosquitoes have built up a resistance to DDT, WHO recommends the use of malathion in these few resistant areas but advises that strict precautionary measures be taken.

Rachel Carson's book *Silent Spring,* published in 1962, lit the anti-DDT torch. Miss Carson stated on page 25 that the malaria-carrying mosquitoes had become resistant to DDT. According to Roy F. Fritz, scientist/entomologist, Vector Biology and Control, WHO, at the present time one percent of the world's population lives in areas where mosquito resistance to DDT has limited its use. This statement was made in a letter to Dr. Thomas Jukes dated March 24, 1971.

Obviously we must never stop our research and our checks on pesticides; however, it would be equally foolhardy to discontinue the use of a pesticide that is demonstrably safe and effective in 99 percent of the world when properly used.

An interesting sidelight on the use of DDT in malaria control is that, in the beginning, it was difficult to convince the people that spraying their homes with DDT was accomplishing anything. Once the walls have been sprayed, the DDT cannot be seen, felt, or smelled; therefore it did not appear to be "strong medicine." Fortunately, the sudden decrease in malaria victims convinced the doubters.

In a Central American country, housewives objected to DDT because after it was used hoards of cockroaches darted out of the walls and ended up in a dead heap in the middle of the floor. However, they decided that cleaning up roaches was a small price to pay for saving lives.

For those who find it hard to relate disease occurring in far-off Ceylon or somewhere halfway around the world to the United States, a little history and the relating of a curious recent development may be useful.

The significance of malaria in our colonial history cannot be overlooked because it was a major hurdle in the development of the American colonies. Directly or indirectly, malaria was one of the most fatal of the colonial diseases and shared with dysentery first place among the colonial infections. Some historians believe that malaria, brought in from Europe, might have accounted for the disappearance of the first Jamestown. It is known that the slaves who were transported from Africa brought malaria with them. Unfortunately, the anophelene mosquitoes carried malaria not only to the slave-owners but also to many other innocent people who protested slavery.

During the American Revolution both armies were plagued by malaria. Dr. L. J. Warshaw, a British army surgeon, noted that the intermittent attacks of malaria put several entire British garrisons out of action and may well have hastened the Yorktown surrender. In response to this comment, Dr. L. C. Duncan, a colonial doctor, stated that the surrender at Yorktown came just in time for the colonials, since our soldiers were also suffering from malaria.

The American pioneers brought malaria to the West Coast. One epidemic of malaria lasted from 1830 to 1833, beginning at Fort Vancouver and moving southward through California. During this epidemic three-quarters of the California Indians were wiped out in a single summer. In these early years malaria was often referred to as the ague. During both the Mexican War and the Civil War, malaria struck soldiers of each side, causing severe illness and many deaths. After both wars, soldiers returning home took malaria with them.

One hundred years ago malaria reached full epidemic proportions in the United States, and even spread into Canada. The largest malaria outbreak occurred in California during the days of the Gold Rush.

As recently as 1941 malaria was endemic in 36 states. In the early days attempts were made to bring malaria under control by draining the swamps around the cities. In 1946 the United States Public Health Service proposed that a five-year National Malaria Eradication Program be started. Congress accepted the proposal. Between 1945 and 1952, 6.5 million houses in thirteen states were sprayed with DDT. As a result of this massive effort all focal points of malaria have been eradicated from the United States. Not one single illness or death occurred as a result of the DDT spray.

Then came the war in Vietnam, a malaria-ridden country.

The American government reported that approximately 10 to 15 percent of our troops in Vietnam were heroin users. One such heroin addict contracted malaria from an anophelene mosquito and then loaned his needle to a buddy. Since the needle is injected into the bloodstream, and addicts do not usually bother to sterilize their needles, that needle became a carrier.

The following is a statement about needle-induced malaria in Kern County, California:

In the three month interval January 15 and April 15, 1971 there have occurred 44 cases of Tertian Malaria in the Bakersfield, Kern County, area. All of these cases have been confirmed by identification of the parasite in the blood. Each of the patients infected was addicted to heroin and gave a history of using common needles with other addicts. The majority were males under 25 years. The source case of the malaria appears to have been a veteran of military service in Viet Nam, who returned to the Bakersfield area about one year ago and who also was a drug user.

Including the 44 noted above, a total of 369 persons have been treated with Chloroquine Phosphate, a drug which is known to completely cure the needle-transmitted form of Tertian Malaria in virtually all cases.

Epidemiologists who have studied this outbreak are fearful that if successful eradication of the disease is not obtained there is considerable danger of a mosquito-borne epidemic during the late summer months in areas such as the Sacramento Valley where the anophelene mosquitoes are known to be prevalent.

> (Signed) OWEN A. KEARNS, M.D.
> Health Officer
> April 23, 1971

The anophelene mosquitoes are not only prevalent in California's Sacramento Valley but throughout most of the United States. Dr. Kearns explained to me in an interview that treatment is given in three doses. He stated that he had no way of knowing if all of the addicts in that area who might be carrying

the malaria parasite had come in for treatment. If one malarious GI is bitten by an anophelene mosquito, an outbreak of
malaria can easily result. Since addicts tend to move around a
great deal, the problem could readily spread to many areas. The
figures cited are for just one county in the most heavily populated state in the Union. Dr. Kearns did not say that *all* cases
had been cured. World War II veterans who have had malaria
know that the disease can and does recur.

TYPHUS   Before the use of DDT, typhus repeatedly swept
across Europe, killing incredible numbers of people. Scientists
estimate that more than 200 million people died of typhus in
Europe and Asia alone prior to the use of DDT. During World
War I, three million Russians died of typhus while millions
more died in the Balkans and Poland.

Tyhpus is transmitted by body lice, which, like mosquitoes,
feed on the blood of animals and human beings. The lice move
from person to person, from animal to animal, or from animal
to human beings, so that typhus spreads rapidly.

The symptoms of typhus are headaches, skin rash, extremely
high fever and delirium. One out of every four persons infected
with typhus dies. With both malaria and typhus, the victim
does not become ill immediately after being bitten. Because
there is an incubation period, immigrants to the United States,
as well as returning GIs, may fall ill from either typhus or
malaria after entering the country.

DDT was first used by the United States Army in February,
1944, to halt an epidemic of typhus fever in Naples, Italy. At
that time the inhabitants of Naples, along with our own service-
men, were dusted with DDT powder to destroy the typhus-
carrying body lice. It was necessary to use 2.2 grams of DDT
per person. DDT can be used in a solution form on clothing and
will keep the clothing free from lice for as long as two months,
even after the clothes are laundered. During World War II,
millions of people throughout Europe were dusted heavily with

DDT by the United States troops. Entire families walked for many miles to reach the American dusting centers and waited in long lines to get the life-saving powder inside their clothes, on their bodies, and in their hair. At many sites the American troops had to work around the clock to give the powder to the people. The DDT dust treatment killed the lice and abruptly halted the spread of typhus in Europe without one single death resulting from DDT.

Today a vaccine against typhus is available, but it is far too costly for most people; consequently countless people still rely on the most satisfactory of protection from typhus—DDT. If DDT were really dangerous and caused harmful side effects, is it reasonable to believe that these people would, after all these years, still be using it voluntarily? In India, Asia, Africa and South America, the vaccine is even less available because of poverty and limited medical facilities. People can and do use DDT. It *is* working. Dr. J. Gordon Edwards, professor of medical entomology, California State University, San José, has said that to stop the production of DDT in the United States, which produces more DDT than any other country in the free world, would be to sentence hundreds of thousands of human beings to certain death from typhus alone.

YELLOW FEVER   Thanks largely to the use of DDT, yellow fever is nearly under control in South America, Central America and some parts of Africa. Yellow fever is carried by the *Aëdes aegypti* mosquito. Like malaria, this dreaded disease is spread by means of a vicious circle.

Although the *Aëdes aegypti* is prevalent in parts of North America, this continent, though once struck by yellow fever, is now free of the disease. Since the mosquito remains, the danger that an outbreak could occur again still exists. If just one mosquito bites one yellow fever victim, the chain is relinked.

Yellow fever prevents the liver from working properly to purify the blood. The victim turns a yellowish color, his temperature rises rapidly, he becomes restless, and his bones ache severely. His temperature drops suddenly, stays down for a few days, then goes up again, and he becomes more violently ill. The mortality rate is high, and death comes quickly.

The French abandoned two attempts to build the Panama Canal because so many of the workers were bitten by the *Aëdes aegypti* mosquito and died. The Americans encountered the same problem. Years were lost draining swamps and clearing underbrush to rid the area of the mosquitoes, a job which DDT could have handled in short order had it been available then.

Like typhus and malaria, yellow fever is not a "contagious" disease. There is a vaccine for yellow fever too, but it is costly and difficult to get to the thousands of people in remote areas. Their only hope is DDT. It kills the mosquitoes, but no ill effects to human beings have ever been reported.

BUBONIC PLAGUE   DDT kills the flea that carries the bacteria that cause bubonic plague. Bubonic plague strikes rapidly, lasts only a few days in each victim, and has a high death rate. The patient experiences chills, fever, and body pain. The lymph glands swell, particularly in the groin, under the arms and in the neck. These lymph swellings are called buboes; hence the name bubonic plague. Black spots of blood appear under the skin, giving the disease the nickname "Black Death."

A rat carries the flea that carries the bacteria that kill the people. If the bacteria get into the bloodstream and thus are carried to the lungs, the patient develops pneumonia-like symptoms (pneumonic plague). When this happens, the disease becomes highly contagious and can be passed from person to person by sputum, droplets in the air, or even by touch. Death usually occurs within two or three days. Seaports are especially vulnerable to bubonic plague because the rats stow away aboard ship and move from port to port. Rat guards are now used on the ropes and hawsers to prevent rats from boarding or leaving ships.

In India alone more than ten million people have died from the plague. In London over 150,000 died of bubonic plague between 1603 and 1665. In the fourteenth century it destroyed one-quarter of the population of Europe. Similar epidemics have swept through other continents. Early in the twentieth century two million people in Egypt and Turkey died of the "Black Death."

According to Dr. J. Gordon Edwards, "There is not enough DDT available now to control the many diseases that could be controlled with its use. It would be utter insanity to ban further DDT production."

Malaria, yellow fever, bubonic plague and typhus are just a few examples of insect-borne diseases which can and are being controlled by DDT.

Ia, another important use of DDT is in control of the tsetse fly, which transmits the parasites of sleeping sickness to man, cattle and horses. As a result of this disease, it was almost impossible to raise livestock in vast areas of Africa before the use of DDT. Sleeping sickness in both man and domestic animals is largely responsible for the slow development of several African countries. Detailed information about other such diseases may be obtained from the United States Public Health Service, Washington, D.C. or WHO, Geneva, Switzerland.

A few years ago Rachel Carson wrote *Silent Spring*, which soared to the top of the best seller list. In this book she repeatedly warned that DDT was a dangerous pesticide. Her book was lyrical, and the material on DDT was far from accurate.

Although in *Silent Spring* Miss Carson made only slight references to some of our vast public health problems, she did write the following (page 187): "Only yesterday mankind lived in the fear of the scourges of smallpox, cholera and plague that once swept nations before them. Now our major concern is no longer with the disease organisms that once were omnipresent; sanitation, better living conditions and drugs have given us a high degree of control over infectious disease."

Her statement was partly true because we have indeed conquered many *infectious* diseases. However, malaria is not one of them, and a death rate of one person every twelve seconds could hardly be considered a "high degree of control." Granted that drugs can cure many diseases after persons have contracted them, but what about that ounce of prevention? Medications are costly and medical facilities are rarely available to people in the vast underdeveloped areas of the world where insect-borne diseases are so prevalent.

Silent Spring is required reading in many of our high schools today. Miss Carson's anti-DDT attitude has given the young reader a *cause célèbre:* "Let's go after DDT and then everything will be fine again."

The National Geographic Society also has dubbed DDT a poisonous pesticide. Its attacks are speculative and only loosely documented. The magazine consistently offers beautiful photographs of Africa, depicting animals, birds and unclad black people in picturesque settings. Their pictures rarely if ever show the ravages of tropical diseases that the African people must endure. In the areas of Africa where DDT is used these dreaded diseases are now rare. Since not one death in Africa has been caused by DDT and several million African lives have been saved by it, and since the *National Geographic* staff takes great delight in photographing Africa, it is difficult to fathom the Society's insistence that DDT is unsafe when the World Health Organization has consistently declared it to be safe.

Unfortunately, much of Africa remains unprotected from malaria and millions of African people still live in malarious areas. Many of these areas are remote and many of the people are nomadic. However, we are making headway, and each year fewer and fewer Africans die from this disease. The program must continue until malaria is completely eradicated.

Seven of the over six thousand documented scientific reports made independently in the past eight years on the alleged hazards of DDT follow. Other references can be found in the list of sources at the end of the book.

1. The President's Science Advisory committee.
2. The Environmental Pollution Panel of the President's Science Advisory Committee.
3. The Committee on Persistent Pesticides of Biology and Agriculture, National Academy of Sciences.
4. The Health, Education and Welfare Commission on Pesticides and their Relationship to Environmental Health, chaired by Dr. Emil Mrak, December, 1969.
5. The Council on Occupational Health.
6. The Council on Environmental and Public Health of the American Medical Association.
7. The United States Congressional Committee on Agriculture, chaired by Congressman Poage, March, 1971.

Not one of these prestigious groups has recommended that DDT be banned. Each group has studied independently the pros and cons, and each has concluded that for many purposes there is no pesticide as effective, or as cheap, or as safe as DDT.

The National Communicable Disease Center, Atlanta, Georgia, in a report released July 25, 1969, stated the following: "For the remaining malarious areas of the world, the banning of DDT production would have catastrophic and tragic consequences which do not seem justified by the available facts. Although DDT has been studied more extensively in man than any other known insecticide, no concrete evidence has been presented that it presently constitutes any health hazard to man. Its use record with regard to human safety is unparalleled in the history of insecticides."

According to a report from WHO, Geneva, Switzerland, issued in 1970, 90 million pounds of DDT are used each year in the battle against malaria alone. In 1968, the United States produced 125 million pounds of DDT, 75 million pounds of which went toward world health. In 1970, 92 percent of the DDT produced in this country was purchased for use abroad.

The Eighth World Health Assembly in 1955 resolved that WHO should sponsor and assist member countries to participate in a global malaria eradication program. Since that time WHO, PAHO (Pan-American Health Organization), and UNICEF (United Nations International Children's Emergency Fund) have substantially assisted 116 countries in which antimalaria programs have been carried on. Originally 146 countries were classified as malarious. Owing to the use of DDT, eradication has been claimed in thirty-six of those, most of which are either islands or more technologically advanced countries, such as the United States and the nations of Western Europe; fifty-three countries are currently engaged in national eradication programs; twenty-seven countries are carrying out significantly large-scale antimalaria efforts; but thirty countries, mostly in Africa, are still without malaria projects.

America *is* vulnerable. Although many diseases have been brought under control in the United States by the use of DDT, including insect-borne encephalitis, plague, typhus, yellow fever, and malaria, it should be remembered that many Americans are frequent travelers to other countries. They, like the GIs, are vulnerable to these diseases, and they can and do bring them back to the United States.

Politics seems to seep into every phase of our living. It is unfortunate that it must influence the production of DDT, but we can hope that our politicians will listen to fact rather than fiction. They might well consider the political consequences of the United States ban on DDT. After using DDT to eradicate malaria from our own land, can we deny other countries the use of the pesticide that helped us? On the other hand, can we expect them to accept and use a pesticide if it has been deemed unsafe for use in our country?

We consider our nation to be the most advanced and affluent in the world. Do we have the right to make a decision to ban DDT when such a decision would have such far-reaching consequences, and would obviously affect the future economic and social well-being of so many nations of the world? Often we hear people say, "Live and let live." Should we not say instead, "Live and HELP live"?

The
Safety
Factor

3 Where our environment is concerned, it is difficult for us as a nation to achieve a reasonable balance between proper concern and unwarranted panic. Instead, we tend to react by extremes, ignoring the fact that nothing in life is entirely without risk.

No pesticide is 100 percent safe nor is any other chemical. Pesticides are used by people who are subject to error. Has DDT ever caused harm to human beings? Not when properly used. There have been a few suicides involving DDT, but even in these cases, DDT was not listed as cause of death. In one case a man bought a quart of a concentrated solution of DDT in kerosene and drank the contents. It was not determined by the doctor whether the DDT, the kerosene, or the combination caused his death. He bought a pesticide and doubtless intended the pesticide to act as the suicide agent; however, the kerosene in that quantity would certainly have caused death.

This is not to say that DDT has no potential for harming the environment. DDT can kill fish and it can kill birds *if it is misused.* There are people who "reason" that if a recommended amount of anything, even cough syrup, is good, twice that amount would be better. Nonsense, of course, but it does happen. The world of pesticides is no exception.

How does DDT affect people? We already know about the life-saving effects of DDT for people who live in danger of insect-borne diseases. We already know that no harm from DDT to the spraymen or to the inhabitants of the sprayed houses has ever been reported. What *does* DDT do to people?

Between 1956 and 1966, when DDT was most widely used in the United States, the average adult ingested up to 0.07 milligram (mg) of DDT per day in his average diet. Later in response to political pressure, the Food and Drug Administration lowered the tolerance level of DDT from 7 to one parts per million (ppm) for most foods where good agricultural practices made this lower level feasible. The average consumption of DDT in the United States in 1971 was only 0.03 mg per adult per day, *not because the tolerance had been lowered* but because by that year DDT was effective against fewer insect pests on food crops owing to resistance problems.

The amount of DDT in food is calculated as milligrams per kilogram (mg/kg). Parts per million (ppm) is a term identical to mg/kg. One part per million is comparable to one jigger of vermouth in a railroad tanker car full of gin, a ratio acclaimed by some gourmets as the perfect martini. Even this minute measurement is not enough for scientists who now have analytical machines that register to parts per billion (ppb) and sometimes to parts per trillion (ppt). If we use the same analogy, one part per trillion would be a single drop of vermouth in 12,-500,000 gallons of gin.

Dr. Francis A. Gunther, of the University of California, Riverside, acclaimed by his peers as the "Father of Pesticide Residue Chemistry," recognized nearly three decades ago the profound effect in food production of the widespread and large-scale use of pesticide chemicals in terms of minute amounts of these chemicals that could persist in foods. He said that some of these effects could be predicted and that others would be unexpected. The accuracy of his 1945 forecast has been borne out.

In reference to finite analytical measurements, Dr. Gunther has said: "I don't think anyone can determine parts per trillion of any chemical, let alone DDT. Our analytical capabilities are just not that reliable. We can determine ppm with a fairly high degree of reliability. In parts-per-billion there is a very high unreliability factor. Parts-per-trillion is a fairy tale.

"People who are trained analysts will not and certainly should not use ppt. Calculators are now widely used and the operator can calculate out to the umpteenth decimal: the machine does all the work for you and you copy down all the numbers. Many analysts are prone to forget the significance of numbers. In the literature we see figures such as 6.7543 ppm. This is ridiculous. Most total residue-analytical methods are not good enough to distinguish accurately between 6.5 and 6.8 ppm; still, because the calculator can do it, the operators run the figures out to four decimal places. People start quoting these figures. This fascination with numbers has caused us to lose our perspective with the total pesticide residue picture."

At one point Dr. Gunther quipped, "Yesterday we looked for little bits of a few things in some things; today we look for less of more things in any thing; tomorrow we will look for nothing in every thing." Another of his all too true observations was this: "It used to be that we could give any analytical chemist a can of paint on Monday morning and by Thursday he would tell us the age of the sample, what vehicle was used, and probably identify the pigments. Nowadays we give the chemist the can of paint on Monday morning and by Thursday he hasn't even decided what method of analysis to use."

Why has DDT become a prime target of the environmentalists? DDT is a complex chlorinated hydrocarbon pesticide with the scientific name *d*ichloro*d*iphenyl*t*richloroethane. The anti-chemical segment of the population has found it an easy target because it has been the single most widely used insecticide in the history of man and because it is an acronym which is easy to remember and to spell. Had they singled out aminotriazole or the atrazine herbicides, the average person would have ignored them.

A tremendous amount of research has been done on DDT because it *is* so widely used, because it is such a fascinating compound, and because it controlled such an immense number of insects. Aspirin is another versatile chemical which has been investigated extensively because it has such a wide usage range.

The instrument now most frequently used to measure DDT is the gas chromatograph. All this instrument does is separate and measure materials in an injected sample. This, in itself, is useless information unless one can identify the materials he has separated. If one were to put a sample of bourbon in the gas chromatograph it would separate that bourbon into twenty-five or thirty different components, one of which would be ethyl alcohol (if the chromatograph were properly adjusted). First, the operator must know at what point this alcohol peaks on the chromatograph. If he does not know this, he cannot identify the *amount* of ethyl alcohol in the sample of bourbon.

There are many kinds of detectors that may be used with the gas chart. For example, there are detectors that will respond only to chlorine atoms. DDT and PCBs (*poly*chlorinated *b* iphenyls, complex organic chemicals used in a variety of industrial processes and products) both contain carbon-chlorine bonds. Prior to the development of more refined analytical techniques in conjunction with gas chromatography, PCBs and DDT, because they peak at virtually the same time after injection, were frequently confused with one another. With improved analytical methodology and equipment, such confusion can be virtually eliminated. The only time a problem should arise is when a given sample contains both. Unfortunately, all too frequently a given sample *does* contain both. This is where great expertise is needed.

There are detectors that respond only to nitrogen or other specific elements such as carbon. One detector responds to any electron-releasing group (oxygen, chlorine, nitrogen, sulfur, phosphorus). This electron-capture detector is most commonly used but it does not tell what it is counting—it only indicates that it is counting something. It is up to the operator to make the identification. There are other detectors that will respond only to phosphorus which are used to detect and measure organophosphates such as parathion.

A gas chromatograph by itself, no matter what detectors are used, does not identify anything. It separates and measures but does *not* identify. It measures signals but it is up to the operator to find out what the signals are. The chromatograph is a superb quantitative tool, however.

By using the gas chromatograph alone, it could never be said with certainty that something is DDT. It could be said that a sample probably is or very likely is DDT but without further analytical study its presence could not be *proved*. The fact that DDT appears at a given interval on a given gas chromatograph does not mean that *only* DDT will appear at that time interval. There are several chemicals out of many thousands of known compounds that would give the same reading, but the chromatograph eliminates the majority of these other chemicals.

Regardless of analytical procedures and the refined technology needed to distinguish between PCBs and DDT, DDT continues to be used as the headliner. The following headline appeared in the *San Jose Mercury,* August, 1971: " 'DDT' Egg Seizure on East Coast." The first paragraph of this story read, "The Food and Drug Administration announced Friday it has seized nearly 7,000 dozen eggs and 45,000 pounds of catfish food because of contamination with a DDT-like substance." The article continued, "The FDA said the seized eggs and catfish feed were tainted with a substance called PCBs—for polychlorinated biphenyls—which have been implicated with causing birth defects and liver damage in test animals." DDT is *never* mentioned again in this Associated Press article; yet it was the first word in the headline. It is true that these two compounds have been confused on the gas chromatograph, but there is no real similarity between the two except for the fact that they both contain carbon chlorine bonds.

When the Monsanto Company, sole supplier of PCBs for the United States, England and Canada, first became aware that this product was causing environmental problems, it voluntarily stopped the sale of all PCBs except for three specific uses: electrical transformers, electrical capacitors, and closed heat transfer systems. According to William Papageorge, a manager in the Environmental Protection Division for Monsanto, "The reason that Monsanto continues to sell PCBs for these three purposes is that the benefits gained in these limited uses far outweigh the problems we might encounter if the material accidently escaped into the environment."

Mr. Papageorge was referring to danger to human beings and property when transformers and heat transfer systems explode or burn. PCBs are used in these devices as a fire-resistant safety fluid. Monsanto should be commended for its attempt to protect our environment. PCBs are *not* subject to any patent. Outside the United States, Canada and Britain, the market is wide open. PCBs are manufactured in Japan, East Germany, West Germany, France, Spain, the USSR, and Czechoslovakia. Products containing PCBs as well as PCBs themselves can be imported from these countries.

Although Monsanto does not sell PCBs to asphalt companies, in many areas of our country PCBs are found in asphalt because asphalt companies buy up refuse such as crankcase oil and transformer fluid to use in their product. There is a vast amount of asphalt on our streets, driveways and highways that contains PCBs and every time it rains, the water washes these PCBs into our sewers and waterways.

There are many ways that PCBs can get into the global environment. One of these is by incineration. Products containing PCBs such as carbonless reproducing paper in business forms, printer's ink, plasticizers, and spent ballasts from fluorescent light fixtures find their way to the city dump where, when burned, they release PCBs in gaseous form. If burning is prohibited and landfill used instead, there is still the problem of runoff after rain. The PCBs can also escape through plant ventilation and exhaust systems; however, elaborate precautions are taken in this country to curtail this. PCBs burn only at very high temperatures. Under normal incineration temperatures, they are vaporized. Through vaporization they are carried into the atmosphere where they collect on minute particles such as dust and pollen and are then returned to the surface of the earth, into rivers, lakes and oceans; hence they contribute to global pollution.

In 1910, thirty-four soil samples were collected in Wisconsin and sealed in glass jars. In 1970 the thirty-four sealed jars were opened and analyzed by means of a gas chromatograph. The analysis indicated the presence of modern chlorinated hydrocarbons (DDT's category) in thirty-two of the thirty-four samples even though not one of these chlorinated hydrocarbons had been developed until thirty years after the samples had been sealed. The reading could not have been showing PCBs because the PCBs were not used until 1929. This is an example of the gas chromatograph indicating the presence of "something," but no one knows what it was detecting. Nevertheless, this report frequently has been quoted out of context to claim that chlorinated hydrocarbons were present.

Using this kind of logic and accuracy, we could make a pretty good story about Teddy Roosevelt and his Rough Riders charging up San Juan Hill in jeeps only to be met at the top by the Spaniards armed with napalm.

The Montrose Chemical Corporation, Torrance, California, is the world's largest producer of DDT. It began its operation in 1947, and from that date to the present time has produced DDT continuously and exclusively. Various forms of DDT are produced by Montrose because different forms are needed depending on the intended usage. For example, they produce technical DDT chips, technical DDT flakes, 100 percent technical DDT dust, and 75 percent technical dust formulated with precipitated silica. In 1967 the plant's production capacity was 7.5 million pounds per month and the average monthly production was six million pounds. At that time, the plant employed between 125 and 165 workers divided into three eight-hour shifts.

In 1967, Dr. E. R. Laws, Jr., M.D., presently with Johns Hopkins University School of Medicine, conducted a study of human beings subjected to extremely high levels of DDT exposure. In addition to the 165 persons employed at Montrose at the time Dr. Laws began his study, a total of 292 people had worked at Montrose six months or more and 806 had worked for fewer than six months, making a total of 1,098 former employees plus 165 current employees. The annual turnover at Montrose is about 35 percent. For comparison, the California State Employment Office reports that the turnover rate in chemical and allied industry for the same metropolitan area was 41.8 percent in 1965. It was obviously impossible to study all of the workers. Dr. Laws' study was based on thirty-five male volunteeers with a work history at Montrose of more than five years. These men were graded by management, their foreman and themselves to determine the average intensity of their contact with DDT. Twenty men had persistently high exposure, twelve men had medium exposure and three low exposure (the three men in the lower exposure category were in contact with considerably more DDT than the general population).

The men voluntarily underwent annual physical examinations which included medical history, laboratory tests of blood and urine, and chest X-rays. They also agreed to donate samples of their blood, urine, and fat for chemical analysis at regular intervals. The only complaint on the part of the volunteers concerned the fat-analysis tests which were done by sampling their body fat with a huge needle in the buttocks.

Before listing the findings and the medical histories of the plant workers, we should note that Montrose Chemical Corporation is located in the heart of Los Angeles County. Another point which might be taken into consideration is the dense population and immense freeway system. Anyone who has ever visited Los Angeles would agree that living conditions in that area are far from ideal. Please note also that some of the individuals had more than one of the conditions listed below:

FINDINGS FROM PAST MEDICAL HISTORIES

Condition	No. of Cases
Acute subdural hematoma followed by post traumatic epilepsy	1
Allergic cutaneous sensitivity to monochlorobenzene	1
Acute labyrinthitis	1
Chronic otitis	1
Chronic sinusitis	2
Pterygia	3
Thyroidectomy for toxic goiter	1
Pneumonia	2
Pulmonary abscess secondary to pneumonia, treated by segmental resection of lung	1
Gunshot wound of chest requiring thoracotomy	1
Influenza	4
Acute bronchitis	3
Asthma with allergic bronchitis	1
Acute myocardial infarction	1
Hypertension	5
Acute multiple embolic episode of undetermined etiology, accompanied by transient hematuria and transient cerebral ischemia in a man with hypertension antedating employment	1
Peptic ulcer	3

Acute urinary tract infection with hematuria	1
Nondiabetic glycosuria with proteinuria	1
Diabetes mellitus	3
Iron deficiency anemia secondary to bleeding peptic ulcer	1

FINDINGS AFTER PHYSICAL EXAMINATION

Condition	No. of Cases
Obesity	3
Congenital deformity, hand and leg	1
Arcus senilis	2
Bilateral pterygia	2
Poor vision, corrected to 20/20 by glasses	6
Poor vision, refusal to wear glasses	1
Minimal hypertensive retinopathy	1
Grade II hypertensive retinopathy	1
Bilateral chronic otitis with bilateral hearing loss	1
Wheezes in lung fields	1
Irregularly irregular pulse (post myocardial infarction)	1
Blood pressure greater than 160/95	3
Soft systolic cardiac murmur	1
Bilateral varicose veins	1
Slack inguinal ring	1
Pilonidal sinus	1

These volunteers were known to have taken in as much as 450 times the FDA tolerance level (0.05 mg) of DDT every work day. The amount of DDT in the fat of these men in all exposed groups ranged from thirty-nine to 128 times the amount found in the general population. The total of all DDT related materials—DDD, DDE and DDA (metabolites or breakdown products of DDT)—was twelve to thirty-two times greater than in the general population. No cancer or blood disease was found in the thirty-five men studied. No cases of clinical poisoning from DDT have ever occurred in the history of Montrose. Dr.

Laws stated, "The average clinical findings of this group of men with heavy exposure to DDT of from nine to nineteen years did not differ significantly from those one might expect from a group of similar age and socio-economic status with no occupational exposure to DDT."

The findings of Dr Laws' 1967 study remain true today. The average family of a Montrose worker includes 4.5 children. In addition to the fact that there has never been a case of DDT poisoning, *not one Montrose employee, in all the company's history, has ever developed cancer either during or after his employment with Montrose regardless of the length of that employment.* No claim is made for DDT as an anticarcinogen; however, the Montrose data, covering a twenty-five-year period, should certainly cause the reader to question the credibility of those who steadfastly maintain that DDT *causes* cancer.

Opponents of DDT allege that this pesticide causes or contributes to a wide variety of diseases in man and animals. Such diseases include cardiovascular disease, cancer, atypical pneumonia, polio, hepatitis and many others. Without exception the cause of any of these diseases was unknown or unproved at the time of the allegation. The charge that DDT caused polio was dropped when Dr. Jonas Salk isolated the polio virus that actually caused polio and developed an effective vaccine against it. In all cases mentioned, the diseases were prevalent before DDT was used. As a result of an indiscriminate attribution of any disease to DDT, research to determine the actual cause might be abandoned or inadequately financed.

Another study of the effects of DDT on human beings was conducted by Wayland J. Hayes, Jr., M.D., Ph.D., formerly with the U.S. Public Health Service and presently with the Vanderbilt University School of Medicine. For twenty-one months (1956–1958), seventeen prison volunteers took up to 35 mg of DDT per person per day under Dr. Hayes' supervision. That is about one thousand times as much as the average

American adult ingested every day during the "DDT Years" (1956–1966). Twenty-two men actually began the experiment but five dropped out because the project interfered with their other assigned duties at the prison and made them unavailable for accurate medical observation. None of the prisoners had any ill effects from DDT.

Fourteen of these volunteers were observed for twenty-five months after they had stopped taking the 35 mg of DDT. None of the fourteen volunteers had any side effects. Dr. Hayes concluded, on the basis of these experiments, that DDT was safe for the general population even at that extremely high dosage. Still the anti-DDT spokesmen persist in labeling the pesticide "dangerous." Why?

For ninety-three days, a California couple, Robert and Louise Loibl, took 10 mg of DDT daily to show that DDT is not dangerous to people. This was approximately the amount that the average person would ingest in an eighty-three-year life span. Mr. Loibl said, "We have undergone biopsies and check-ups and there have been no deleterious effects. In fact, we didn't have any illnesses, not even a sniffle." It should be noted that the Loibls were strongly advised by medical doctors not to conduct such an experiment. It is never wise to diet or to take any unusual or even common non-food chemical without medical advice. Such haphazard experimentation, though doubtless done with the best of intentions, could have done considerable harm. It is true that much larger amounts were ingested in Dr. Hayes' experiment and in the Montrose study with no ill effects. Nevertheless, scientific experimentation should be left to the scientists. The Loibl "experiment" was not science and it proved absolutely nothing.

Why do the anti-DDT people oppose the use of DDT? Who are they, and what reasons do they give? In the Sierra Club's magazine, the *Argonaut,* January 1970, the first sentence on the first page said in boldface type and capitals: PREPARE TO

MEET THY DOOM. The article, written by Don Serebrakian and Kristin Camitta, continued, "Not very pleasant in the way of a New Year's wish. Many of us have seen this sign at one time or another and may have suppressed a smile. But what if the man carrying the sign had tugged at your sleeve and offered irrefutable facts to back this somber prophecy? Leaders of the scientific community foresee the end of life on earth within fifty to one hundred years." Which pesticide is pointed out as the main offender? DDT, of course.

One advertisement was so outlandish in its claims that it was entered into the Transcript of Hearing Testimony for the Federal Pesticide Control Act of 1971 (Poage Hearings) in the hope that government action would be taken against such advertising. The National Audubon Society had purchased a full-page advertisement in the *Saturday Review,* November 7, 1970, that linked cancer and DDT with a picture of an aerosol can showing the skull and crossbones and the letters DDT on the can. Incidentally, products containing DDT have never been required to carry the poison symbol. The complete text of the advertisement is cited directly from pp. 751–752 of this transcript:

ANYONE WHO READ THE RECENT PRESS REPORTS ON DDT AND CANCER IS PROBABLY ASKING, IS ANYONE DOING ANYTHING ABOUT IT?

Yes.

Up to now the National Audubon Society has been a primary force in a relentless battle to ban DDT.

But from now on it has to be you and the National Audubon Society.

Here's why.

THE WORLD'S NO. 1 POLLUTANT BECOMES THE MAIN THREAT TO SURVIVAL.

The U. S. chemical industry is producing 140 million pounds of DDT this year. For which the American public will continue to pay a high price. In the form of a ravaged ecosystem and extinct wildlife.

More and more poisoned food and water. And the threat of cancerous tumors and birth defects.

WHILE WASHINGTON DRAGS ITS FEET, THE PRO-POISON PRESSURE GROUPS PREPARE FOR RENEWED ASSAULTS ON THE ENVIRONMENT.

The opposition to ecological sanity is formidable. The agricultural community, Woolen mills and foresters. The chemical industry. Which, unfortunately, continues to be supported by homeowners, gardeners, and outdoorsmen who should be reading labels and boycotting poisons that endanger their own families.

Government standards allow a certain amount of DDT contamination in your food. The trouble is, there's not enough money or manpower to police all the excesses resulting from overspraying by overzealous farmers. So, even though you pay taxes to have your foods inspected, you get poisoned anyway.

THERE ARE ALTERNATIVES TO SPRAYING THE AIR WITH A CANCER AGENT.

To name a few: safer pesticides like desiccants, pyrethrins, malathion, and nicotine sulphate. And crop rotation. And water control—which removes stagnant water where mosquitoes breed.

INVEST $IO. HELP REVERSE THE AGE OF SELF-DESTRUCTION.

Become a member of Audubon.

Your dues will help us fight the insecticide/homicide battle in Washington, in the courts and in the State legislatures. And your involvement as an environmental activist in any one of our 200 local chapters will also help our crusade to curb population growth, the misuse of technology and many other environmental ills.

As a member, you'll also receive Audubon magazine, the most respected and beautiful conservation magazine.

Our work so far against DDT has given your body a reprieve.

Now help us give you a pardon.

It must be a great comfort to the American Medical Association and the American Cancer Society to know that The National Audubon Society had everything under control way back in 1970.

Another organization "using" DDT propaganda is the Ecological Food Society. It purchased a full-page advertisement in the May, 1971, issue of *Sunset* magazine, which showed a picture of an apple with the caption, "The DDT-less apple—Yours for 9¢." The first paragraph of the ad stated, "This is not a scare ad. If you can read, let alone smell, taste or breathe, you're probably scared enough because you know that a flood of poisonous chemicals has invaded your family's diet, robbing your food of its flavor and wholesomeness, robbing you of your health and perhaps your life as well." They are selling a product called Earth-1, a nonsynthetic cleaner for "things and people." They offer to give you the product free of charge if you become a member of the Ecological Food Society and send them $3.50. What does a DDT-less apple have to do with a synthetic food cleaner? It appears that all one has to do to capture the public's eye is mention DDT.

The fact that something is printed in a newspaper, book, or periodical does not necessarily make it true. An Associated Press story from Phoenix, Arizona, carried the headline, "DDT May Cause Sterility." In this article, Dr. William Heinrichs, an associate professor of obstetrics and gynecology, University of Washington, discussed his current research project. Dr. Heinrichs had injected 3 mg of DDT (one hundred times the amount normally taken in by *adult human beings* each day) into several *newborn* female rats. If we compare the amount of DDT he gave the baby rats, in proportion to size, with the amount human beings ingest, in proportion to size, we see that the quantity given to the rats would be approximately 200,000 times the norm on a body weight basis for a human being. Animals and people are not normally *injected* with DDT. They *ingest* minute amounts of DDT. It is remarkable that he did not kill most of the baby rats. Had he injected water or air into their bodies he probably would have. Dr. Heinrichs found that when these rats reached puberty they had irregular menstrual cycles. When he killed them to do a post-mortem, he found that their ovaries contained tiny growths. The doctor said, "DDT was found to induce in the rats a form of sterility which bears many similarities to relatively common and apparently recently increasing abnormalities among women called the polycystic ovary syndrome." He claimed that many of these women failed to develop egg-producing menstrual cycles and that they usually consult with a physician in regard to irregular menses, infertility, or signs of masculinization such as facial or body hair growth or acne. Yet women have had all of these problems centuries before DDT. If DDT actually causes sterility, why has the population of those nations with the highest DDT usage increased greatly rather than decreased?

In an editorial under the title "Non-Science" in the *British Review of Food Nutrition and Food Science,* the editor discussed at length animal experimentation and set down some "rules" to be followed by scientists who are somewhat less than objective in designing and drawing conclusions from animal experiments:

Rule 1—Give heroic doses The results of animal tests are applied to man, so play it safe. Determine the effects of massive doses and conclude that ordinary use levels may have the same effect in man, only it may take longer or may not be recognized as causally related.

Rule 2—Inject, instead of feed Ignore the factors involved in oral administration, such as rate and extent of absorption, gastrointestinal changes, systemic biotransformations, blood and tissue levels, and the normal excretory routes. Instead pump a large dose under the skin or into the muscle, vein, or abdomen of an animal (one will do), then conclude that the effect observed may occur in women who are—or who may not know they are—pregnant.

Rule 3—Dose fasting animals To get best results give large doses of the chemical substances to animals on an empty stomach. Forget the fact that in normal use, food chemicals are consumed in food, and generally in only low concentration. (By following this rule, common salt will be found to be more lethal than most additives.)

Rule 4—Feed a single food as the total diet Disregard the fact that animals, just like people, need balanced diets with adequate proteins, calories, vitamins, minerals, etc. See how far they can get on only a single food regardless of how nutritionally incomplete it may be. Never mind that the only single complete food that nature has provided is breast milk for infants, that ordinary bread or milk (or yogurt) will not sustain normal health without added supplementation.

Rule 5—Use exotic test animals   It does not matter that in respect to similarity of their physiology or biochemistry to that of man, little is known about such species as the gecko, the ferret, the quail, the hairless mouse, or even the hamster. What's bad for a guppy or mosquito may be bad for man.

Rule 6—Challenge the newborn   Get them before they have adapted to their new environment, and built up their normal defenses. Newborn animals may tell us about the wisdom of injecting chemicals into day-old infants, even if they can't indicate how older children or adults will respond.

Rule 7—Sound the alarm   Don't wait for verification of your findings. Reporters are waiting for scoops and headline stories. They haven't the time to do their homework or check the validity of your conclusions or speculations. Any suggestion that a common food or environmental chemical may cause heart disease, cancer, miscarriage, or premature baldness will spur industry and government (us taxpayers) on to support more research and (who knows) a big grant for you.

The so-called "organophosphates" (from organic phosphates) are important and widely used pesticides which contain carbon and phosphorus, hydrogen, oxygen, and often sulfur and nitrogen. They are generally expensive, short-lived, highly toxic and readily absorbed through the skin. A rule of thumb: If the name of the pesticide ends in "thion" it is an organophosphate (e.g. malathion and parathion).

The following statement regarding organophosphates was issued by the California Department of Public Health in 1967: "Given their high toxicity and frequency of use, it is not surprising that the organic phosphates are today the main killers and cripplers of agricultural workers in California. In 1966 the organic phosphates were deemed to be responsible in 74 percent of the serious pesticide injuries reported to California doctors. Forty-four percent of all the persons who have been killed by pesticides while they have been working have been killed by organic phosphates." No worker has ever been killed or injured by DDT, which is *not* an organic phosphate. DDT is a chlorinated hydrocarbon, or organochlorine pesticide.

The following appeared in an editorial in the *New York Times* on August 23, 1970, headlined "No Improvement":

"The tragic wave of parathion poisonings in the South, mostly of children, raises some extremely pressing questions for the government; if the use of DDT is to be discouraged or banned altogether—as we emphatically believe it should be— doesn't the United States owe its cotton, tobacco and fruit growers the most authoritative advice on safe pesticides? Doesn't it in fact owe them what ever financial aid they may need to make up the difference in cost?

"The danger of hard pesticides like DDT is now too well established to warrant any defense of their continued use. Though less immediately toxic than parathion and related chemicals, DDT increases its lethal touch from plant to insect to bird or fish to mammal. Building up in fatty tissue it has such lingering power that two-thirds of the DDT man has sprayed since it came into wide use in World War II is believed to be still active.

"Yet farmers are often hard put to it to make a sudden change. Most of the substitutes, while far quicker to dissipate, are much more toxic in the first few days after use. The cost, moreover, may range from twice to six times as much as DDT, making the switch sometimes as painful economically as it is physically."

It would be painful in more than economic and physical ways to pay more money for a pesticide that might well kill your child. The *New York Times* should recognize that before discarding the only *feasible* and *safe* pesticide, DDT, there should be proven alternatives immediately available in the amounts needed. The charges of the lethal dangers of DDT, in the second paragraph of the article, will be refuted in a later chapter.

In North Carolina in 1966, after tobacco farmers began using parathion to replace DDT, three deaths and thirty injuries were traced to parathion.

Dr. Charles Wurster, a founder of the Environmental Defense Fund (EDF) and a leading opponent of DDT, was interviewed by James Miller for an article in the October 1969 issue of *Reader's Digest*. Mr. Miller asked Dr. Wurster the question, "What do you regard as the most important of these non-debatable facts about DDT?" Dr. Wurster answered, "First, DDT is an extremely long lasting chemical. The insecticide parathion is far deadlier than DDT, but on an environmental scale it's much safer because it loses its potency in just a few days. If a man applying parathion to an apple orchard gets a drop of it on his skin it could kill him; but a few weeks later we could safely eat the apples he sprayed."

The government of Honduras announced on May 14, 1971, that thirty-seven soldiers of the Presidential Honor Guard, all of whom were young draftees, had died after eating food that had been sprayed with the organophosphate pesticide parathion. Three other young soldiers were in critical condition from the same pesticide poisoning.

Sunset magazine, June, 1971, carried a full-page advertisement for another organophosphate, malathion: "In fact malathion is recommended by two groups actively seeking pesticide control—the National Audubon Society and the Environmental Defense Fund." In the last paragraph of the advertisement, the reader is urged to write to the American Cyanamid Company to learn how malathion can help make life more livable around the home and garden.

It is true that malathion is the safest of the common organophosphate pesticides, but it is not as safe as DDT. Since DDT cannot be purchased in all states, malathion *is* the safest available home-garden pesticide; however anyone who uses this product should read the label carefully. The label on the Ortho bottle containing 50 percent malathion solution reads:

Caution. Harmful if swallowed. Avoid prolonged breathing of spray mist. Avoid prolonged or repeated contact with skin. Malathion is a cholinesterase inhibitor and can cause symptoms similar to those caused by other organic phosphate compounds. Do not use, pour, spill or store near heat or open flame. Keep children and pets off treated areas until this material is worked into the soil and grass is dry. This product will kill fish. Keep out of any body of water.

Malathion can be found in many stores within easy reach of young children. The product is in glass bottles and contains up to 50 percent malathion plus carriers.

The full impact of this warning, though scientifically accurate, may not be recognized by the gardener who is not aware of the symptoms "caused by other organic phosphate compounds" such as parathion. The gardener also may be unaware of the fact that a cholinesterase inhibitor, depending upon the amount and the size of the individual, can cause muscular spasms resulting in respiratory and cardiac failure.

Malathion has been used in concentrations of one percent, 5 percent and 10 percent mixed with talcum powder for controlling some strains of body lice that have become resistant to DDT. This dust was developed by Dr. Wayland Hayes, Jr., and at this low concentration, no harmful effects to human beings have been detected.

If malathion were to be used on a world-wide basis instead of DDT, several hundred thousand trained spraymen would be needed. In the use of DDT, only the supervisors need to be trained. Even now with DDT, we are unable to spray the vital areas as often as needed, owing to lack of personnel. To use malathion, the spraymen would have to wear special clothing for the *monthly* spraying. You cannot let an untrained Bushman spray malathion and say to him, "It doesn't matter if you squirt yourself." Prolonged exposure would kill him.

By contrast, DDT is non-toxic to the spraymen who apply

it and it need be applied only every four to six months. It can be safely applied to clothing, houses, people, and to crops. It should not be sprayed directly on prepared foodstuffs because it often contains a petroleum carrier. People applying it do not have to be trained. An African Bushman can apply DDT just by pumping a handle and aiming a nozzle. If he squirts himself it really doesn't matter. It will not hurt him. He should bathe to get the oil off his skin, but even if he doesn't, there would be no harmful effect except a bad odor.

In addition to replacing DDT with malathion, which costs five times more in wholesale quantities, we must have trained personnel to apply it. Where are we going to get such men and women, and who will train them? They would have to wear expensive protective clothing—coveralls, boots, gloves, goggles, and respirators. Since this spray would have to be applied four to six times as often as DDT, Dr. Gunther estimates that we would have a total increased cost factor of at least sixty. WHO spends approximately one hundred million dollars annually in malaria control *alone*. If malaria control is going to cost sixty times as much where will the additional six *billion* dolars come from?

The time will come when DDT will be replaced, as most pesticides are. It must not be replaced or banned until a safe substitute is available in both quality and quantity at a price all nations can afford. Safe methods of application are vital.

The
Mother's Milk
Controversy

4 The entire controversy over DDT and mother's milk has been hashed and rehashed and exaggerated far beyond reasonable proportions. It must be discussed again here, not because it is significant to either man or his environment, but because it has been used and misused by the anti-DDT factions as a dramatic attention getter.

The executive director of the newly formed Environmental Defense Fund (EDF), Roderick A. Cameron, spent 5,000 dollars of the organization's remaining 23,000 dollars to purchase an advertisement in the *New York Times* on March 29, 1970. Headlined "Is Mothers' Milk Fit for Human Consumption?," the advertisement solicited new members at 10 dollars for a basic membership. After this profitable advertisement the EDF launched a direct mail campaign, and by the end of 1971 had some 25,000 members. According to an Associated Press story in November, 1971, the Environmental Defense Fund expanded with the establishment of a West Coast branch office. At this writing, the EDF is operating on a 750,000-dollar annual budget.

Victor J. Yannacone, Jr., an attorney, provided the impetus

for the Environmental Defense Fund in 1966 when he filed suit on behalf of his wife Carol and "all other people of Suffolk County, Long Island," against the Suffolk Mosquito Control Commission, whose dumping of DDT was alleged to be the cause of a fish kill in a lake near Mrs. Yannacone's childhood home. This suit brought together Mr. Yannacone, Dr. Charles F. Wurster, an associate professor at the State University of New York, Stony Brook, and Dr. George M. Woodwell, an employee of Brookhaven National Laboratories, to found the EDF in 1967. This organization is aided financially by the National Audubon Society, indirectly by the Ford Foundation, and the Citizens' National Resources Association, Wisconsin.

Mr. Yannacone was placed on a 5,000 dollars per month retainer (in addition to his regular practice). His written agreement with the EDF included the provision that when work took him more than one hundred miles from home for more than three days at a time, he was entitled to take his wife and son, and that accommodations for travel, meals, and lodging were to be first class.

The following appeared in *Science,* December 26, 1969: "Last winter Charles F. Wurster, Jr., one of EDF's founders and chairman of its Scientist's Advisory Committee, spoke warmly of Yannacone. 'Vic really thinks he can save the world,' Wurster said. 'He's a brilliant guy. If you aim him in the right direction, he'll raise hell.' "

In a letter to Congressman W. R. Poage, March 9, 1971, which was entered into the Poage Hearings, Dr. Wurster conversely stated, "It was in part because Mr. Yannacone lost touch with reality that he was dismissed by EDF, and his remarks of May, 1970 indicate that his inability to separate fact from fiction has accelerated."

Something serious happened within the framework of the EDF in 1969. Since it was the mother's milk advertisement purchased by this organization that contributed so greatly to the scare regarding DDT, it will not be amiss to take a close look at the feuding and the mental gymnastics of EDF's founding fathers.

A section of the Transcript of Hearing Testimony from the Pesticide Control Act of 1971 is worth noting in light of the unusual answers given by Edward L. Rogers (successor to Mr. Yannacone as general counsel for the EDF).* In this section Mr. Rogers is responding to the questions of Congressmen T. G. Abernethy and J. R. Rarick, members of the House Committee on Agriculture. From pages 266, 267, and 268 of this transcript:

Mr. Rarick.   This Dr. Charles Wurster that you identified, is he the only Dr. Wurster that has been associated with your legal, your Environmental Fund, or do you know of another Dr. Wurster?

Mr. Rogers.   Dr. Wurster's former wife is also a biologist and has done work in this field. I think they did some paper together, which I think is—I think it is Doris Wurster, I am sorry, it is not in this, I do not find it here. I am quite sure that is correct.

Mr. Rarick.   Well, this, would be then the same Dr. Wurster whom Mr. Yannacone, in a speech on May 20, 1970, at the Public Relations Luncheon Group of the Union League Club in New York, described at a press conference as having said this:

A reporter asked the same Dr. Wurster whether or not the ban on the use of DDT wouldn't encourage further use of very toxic materials, including nerve gas derivatives, and he said, "probably."

The Reporter than asked him if these organo phosphates did not have a long record of killing people. And Dr. Wurster said "so what? People are the cause of all the problems. We have too many of them. We need to get rid of some of them and this is as good a way as any."

* A copy of the transcript of this Hearing (Serial No. 92-A) may be purchased from the Superintendent of Documents, U. S. Government Printing Office, Washington, D. C. 20402. Price $3.75.

Is this the same Dr. Wurster that you are later to give us an affidavit from?

Mr. Rogers.   I would say probably not. In fact, I would say very emphatically that I would doubt very much that it is the same Dr. Wurster. I think that perhaps it is a figment of someone's imagination somewhere.

Mr. Rarick.   Do you know John Yannacone?

Mr. Rogers.   Yes, I do know Mr. Yannacone.

Mr. Rarick.   He is one of the founders?

Mr. Rogers.   He is no longer with the Environmental Defense Fund, for very good reasons.

Mr. Rarick.   Would this be the same Dr. Wurster who said when asked the question, "Doctor, how do you square this killing of people with the mere loss of some birds?" And this very eminent, well-meaning scientist said:

It doesn't really make a lot of difference, because the organo phosphate acts locally and only kills farm workers and most of them are Mexicans and Negroes.

Would this be the same Dr. Wurster whose affidavit you promise to later supply us?

Mr. Rogers.   I think this is a very serious matter that you are bringing up here, and I do not know what the rules of the committee are, but we are allowing to be read into the record what is notorious hearsay, without the advantage of any right of rebuttal or examination of the people who reportedly made these statements. If there is any procedure for expunging this in the record, I would ask that it be done.

Mr. Rarick.   I am going to ask unanimous consent, Mr. Chairman, that all portions of the speech given by Victor Yannacone, Jr., attorney at law, Patchogue, N.Y., given at the Public Relations Luncheon Group of the Union League Club, New York, May 20, 1970, limited to those sections relating to the Environmental Defense Fund and the various suits and interworkings on the banning of DDT, be placed in the transcript on this day following this gentleman's testimony.

Mr. Rogers.   Mr. Chairman, may I address myself to that point for a minute?

Mr. Abernethy.   He asked the committee's permission to do something. I think you can address yourself to it, but I do not think you have any control over it.

Mr. Rogers.   I realize that; I would like to clarify the point that Mr. Yannacone is no longer with the Environmental Defense Fund. At the time he made this speech, he was not with the EDF. The party [*sic*] was not altogether amiable.

Mr. Abernethy.   Did he make the speech?

Mr. Rogers.   I have no idea.

Mr. Rarick.   Mr. Chairman, in this speech, Mr. Yannacone indicates why he is no longer with the EDF. He said it is because of Dr. Wurster.

(The speech by Mr. Victor J. Yannacone, Jr., referred to above may be found in the files of the committee.)

(In connection with the foregoing quotes of Dr. Charles Wurster, the following letter was received from Dr. Wurster:)

March 9, 1971
CONGRESSMAN W. R. POAGE Chairman, House Agriculture Committee, House Office Building

Dear Congressman Poage:

It has come to my attention that certain remarks attributed to me by Mr. Victor J. Yannacone, Jr., in May 1970 have been inserted into the record of your hearings on pesticides during the testimony of Edward Lee Rogers of the Environmental Defense Fund.

I wish to deny all of the statements of Mr. Yannacone. His remarks about me, attributable to me, and about other trustees of EDF are purely fantasy and bear no resemblance to the truth. It was in part because Mr. Yannacone lost touch with reality that he was dismissed by EDF, and his remarks of May 1970 indicate that his inability to separate fact from fiction has accelerated.

I respectfully request that my denial of any truth to Mr. Yannacone's remarks be made part of the record of these hearings.

Sincerely,

CHARLES F. WURSTER

Associate Professor of Environmental Sciences,

Marine Sciences Research Center,

State University of New York,

Stony Brook, N. Y.

This is the same Dr. Wurster who, in the *Reader's Digest* reference made earlier, argued for the use of the deadly insecticide parathion despite the fact that by his own admission a single drop of parathion on the skin of a man applying this insecticide would kill him.

In a May, 1970 article in *Science Review*, "DDT in Mothers' Milk," Dr. Wurster made two conflicting statements. On page 58 of this article he wrote, "The biological significance of the DDT residues we ingest is difficult to interpret. We know that DDT does not kill people outright, even at far higher dosages than we or nursing babies take in. Nor are there obvious overt symptoms of DDT poisoning." On page 59 of the *same* article, Dr. Wurster stated, "The presence of DDT in human milk is not just a problem for nursing mothers. With the profound impact on the biota of our planet, DDT has become a threat to us all." He went on to say, "If all environmental problems require a quarter century of scientific research, controversy and legal battle before effective action reduces the contaminating biological damage, as has been the case with DDT, it may be too late for mankind. Let us therefore take action, not only on a local level, but internationally as well, through such a body as the United Nations, to guarantee the survival of the global ecosystem."

At almost the same time in *Highlights,* 1970, Mr. Yannacone stated, "Any law systematically banning the use, sale, manufacture or distribution of DDT in your state, county, city, or even the United States, without at the same time establishing an ecologically sophisticated pesticide regulation program, is a bad law. It won't satisfy anyone very long and will permanently polarize agriculture and conservation to such an extent that common problems can no longer be considered in internal discourse."

On October 8, 1969, the following headline appeared in the *Seattle Post Intelligencer:* "Mothers Want DDT Ban To Be Made Nationwide." Under the headline appeared the following copy, "Charging that DDT may cause cancer, a group of California mothers yesterday petitioned the Food and Drug Administration (FDA) to ban the pesticide nationwide. The petition presented by lawyers for five pregnant and nursing mothers said DDT spray on food might be polluting mothers' milk and exposing breast-fed babies to increased danger of cancer."

The actual words of the petition to the FDA were, "Recent scientific investigations have provided unimpeachable evidence that DDT causes cancer in animals and provides very strong indication that it produces cancer in man."

Nursing mothers, frightened by this charge, were promptly reassured by California Public Health Director Dr. Louis W. Saylor, who issued the following statement: "We have seen no increase in childhood cancer since the presence of DDT in milk and other food became common so many years ago. Our studies show that breast-fed babies are as healthy as bottle-fed babies. Furthermore, there are psychological and immunizing advantages to breast feeding."

Within the mothers' petition was the demand that the FDA "ban any food containing residues of a substance that causes cancer in animals." Had this petition been accepted as submitted, the FDA would have been forced to ban *all* foods because *all* foods contain substances which can be shown to cause cancer in experimental animals, given the appropriate dose and the receptive animal.

Dr. Jay M. Arena, professor of pediatrics, Duke University School of Medicine, placed the concern over mother's milk in its proper perspective in his 1970 article in *Nutrition Today,* "Contamination of the Ideal Food." Dr. Arena wrote that nursing mothers are exposed to environmental toxicants just like anyone else. He pointed out that DDT is present in all human milk to a slightly greater degree than in cows' milk, a fact that has been known for years. Dr. Arena added, "Most statements about the harmfulness of DDT in breast milk contain half-truths and misrepresentation." He also stated that a review of all the available literature on the subject indicates a steady decline of DDT in both mother's milk and cows' milk over the years.

DDT has been present in both cows' and mother's milk since the time of DDT's original use. There has never been a case of death or injury reported as a result of the presence of DDT in either.

In 1959, the first administrative guideline for DDT in cows' milk was set at 0.1 ppm (0.1 gram for 250 gallons). This number was adopted by the FDA based on residue analytical capabilities *at that time,* not on any toxicity considerations. At that time analysts could easily detect 6.8 ppm of DDT (if present) in most raw foods which, incidentally had a tolerance level of 7 ppm (established by the FDA), but could not reliably detect a residue to the second decimal point such as 7.15 ppm. *There was no alarm and there were no problems related to DDT in milk.* Then improved analytical methods were developed that *could* detect to 0.01 ppm and lower. In 1965, in response to pressure, a tolerance level of 0.05 ppm was *recommended* for DDT in cows' milk. This recommendation did not become *law* until November 25, 1971. When it was pointed out that the previously recommended tolerance level for the whole milk they had been drinking was nearly twice the new level, people panicked. This 0.05 ppm DDT tolerance level in milk is toxicologically and pharmacologically (which includes physiological effects) insignificant. We must continue to be alert; however, we must not allow an excessive fascination with analytical numbers to interfere with objectivity.

The following is a statement issued in 1966 by the Committee on Appropriations of the U. S. House of Representatives:

FDA regards cows' milk as a special food because it is regularly fed to sick persons, the aged, and infants. However, representatives of AMA stated that, contrary to the opinion of the FDA, invalids and the aged are no longer put on all-milk diets, except under most unusual circumstances. The Food and Nutrition Board of the National Academy of Sciences has stated that cows' milk should not be treated as an exceptional food. Representatives of AMA stated that there is no medical reason for singling out cows' milk as an agricultural product for which pesticide residues are not permitted. They stated that babies who are bottle-fed with cows' milk are, nowadays, put on a diet containing other foods when they are still very young, and these foods often include processed baby foods for which finite limits for content of pesticide residues have been set under FDA regulations. Therefore, the baby is not permitted to ingest pesticide residues when he is fed cows' milk, but may take them in other foods.

In contrast to bottle-fed babies receiving cows' milk, breast-fed babies are ingesting pesticide residues contained in the mother's milk because she ingests foods that legally contain pesticide residues. Studies were made by PHS at the Taft Sanitary Engineering Center in Cincinnati, Ohio, and by the Bureau of Occupational Health, State of California, Berkeley, California, in which it was determined that samples of mother's milk contained amounts of pesticide residues far in excess of the pesticide residues permitted by FDA in cows' milk. Further, the study concluded that there were no ill effects on the babies as a result of the pesticide residue in the mother's milk.

Dr. Robert White-Stevens of the New Jersey Agricultural Experiment Station, Rutgers University, determined that it would be necessary for a mother to give an eight-pound baby from 70 to 233 gallons of milk as an exclusive source of nourishment before the total amount of DDT equalled what the developing child received through the mother's placenta. This would only occur with 100 percent absorption and no metabolism or elimination of DDT, which is physiologically impossible.

For the purpose of comparison, it is interesting to note that the intake of DDT in India, as reported in 1967 by Dr. Wayland J. Hayes, Jr., was twenty times as high as in the United States that year. As a result of this, the suckling infant in India would have ingested a far greater amount of DDT than the American infant. Yet the infant mortality rate continues to decrease in India.

The concept that milk is free from so-called toxic substances is an illusion that does not fit the facts of science. To reiterate, "Everything is poison, yet nothing is poison," depending upon the quantity as well as the identity of the substance. If a mother has been smoking, her milk will contain nicotine, a well known poison. Recently a young mother questioned me at length on whether it would be safe for her to nurse her child. She feared that the DDT in her milk might damage the infant. I talked with the young woman for an hour. During this time, she chain-smoked seven cigarettes, drank two highballs (bourbon and water), and took one tranquilizer which she said contained 400 mg of meprobamate.

It is a known fact that most drugs are passed on to the suckling infant through the mother's milk. Milk has always contained minute amounts of arsenic, radioactive carbon, and large amounts of cholesterol, and no one is unduly alarmed about that, nor should they be.

A San Francisco doctor recently delivered a young heroin addict of her infant son. Unfortunately, the baby was born an addict. The doctor had explained to the mother that this would happen. Yet her first question after the baby was born was, "Will the DDT in my milk harm my baby?"

According to Dr. Walter Ebeling, professor of entomology, UCLA, "When DDT usage was at its height in the United

States between 1956 and 1966, the average concentration in mother's milk was 0.13 ppm. Along with the decline in DDT usage, the concentration in mother's milk has declined to a maximum of between 0.03 and 0.04 ppm." The difference between 0.13 ppm and 0.03 ppm is minuscule at these low levels, and is indicative of the need for a more rational perspective regarding quantitative analysis in toxilogical applications.

In the hopes of achieving a more rational perspective on the subject, Dr. Emil Mrak, Chairman of the HEW Commission on Pesticides, has stated, "We need a study of our *total* food habits. We're living in an era of food McCarthyism. We're afraid of everything. We worry about everything. Yet few scientists and almost none of the general public are sufficiently informed to identify the actual hazards in food or even to know if there is cause to worry. The highest form of malnutrition—overconsumption of fatty food—affects millions of Americans." Dr. Mrak believes that the overconsumption of fatty foods can be more harmful than additives and pesticides that are found in foods.

Dr. Alice Ottoboni, research toxicologist for the California Health Department, has said that although the amount of DDT in mother's milk is somewhat higher than the amount found in cows' milk, the medical profession has specifically recommended that breast feeding not be abandoned. Dr. Ottoboni noted that the health of breast-fed children is viewed by the medical profession as generally better than those raised on cows' milk.

Regrettably, medical facts and solid professional opinion have had little impact on those who choose to use mother's milk as a propaganda tool. The Audubon Society's *Nature Trails* contained an excerpt from the June 15, 1970, *Conservation News*, in which Ed Chaney of the National Wildlife Federation wrote:

A five day old human being lies asleep in the other room. His name is Eric. His tiny, wiggly, red body contains DDT passed on to him from his mother's placenta. And every time he sucks the swollen breasts he gets more DDT than is allowed in cows' milk at the supermarket. Be objective? Forget it. Objective is for fenceposts. How can you be objective in the face of a global insanity that is DDT?

. . . . Who knows the name of one DDT company president? Just what exactly was the name of the man who reportedly decided agriculture would use the public's money to go down kicking and screaming in defense of DDT? A man did it. Not the Agriculture Department. Not the buildings, desks, and organization charts. Nor The System, but a man who puts his pants on one leg at a time. A man who had the responsibility, opportunity, and authority to end the environmental madness that is DDT. And he copped out.

Objectivity? At least we will have our vengeance, Eric and I. Now that conservation is considered relevant, the political historians will ferret out that man and deal with him unkindly. And most shamelessly delightful, he will have to deal with himself.

"Breast milk is the ideal food for infants," according to Dr. Arena. "It is the one food that cannot be improved upon. The only way breast milk can be rendered less than the perfect baby food is for the mother to take drugs that pass into her milk." Dr. Arena's chart, "Drugs Excreted in Human Milk," is self-explanatory. Dr. Arena has explained that if a mother can produce enough milk for her baby, infant malnutrition is virtually impossible. Even amidst the worst poverty conditions, the suckling baby will be well nourished. The most prevalent disease of protein-calorie malnutrition in the world is *kwashiorkor,* a name given the condition by the Ga tribe of Ghana, who first used the term to describe the sickness (malnutrition) the child gets when it is weaned.

Rules, regulations, and laws may be passed at the city, county, state, and federal level; but individuals must always apply the rule of reason. For a mother to refrain from nursing her baby because she fears the effects of DDT would be to deprive both mother and child of the psychological advantages of breast feeding and would also deprive the child of the physiological and immunizing advantages.

Alcohol
Allergens
Ambenonium chloride (Mytelase)
Aminophylline (theophylline with ethylenediamine)
Amphetamines
  Amphetamine sulfate (Benzedrine and numerous other trade names) and other salts of amphetamine
  Dextroamphetamine sulfate (Dexedrine and numerous other trade names) and other salts of d-amphetamine
Analgesics (non-narcotic)
  Acetaminophen (Amdil, Anelix, Apamide, Elixodyne, Febrolin, Fendon, Lestemp, Lyteca syrup, Metalid, Nacetyl, Nebs, Tempra, Tylenol)
  Aspirin
  Dextropropoxyphene hydrochloride (Darvon)
  Phenacetin
  Sodium salicylate
  Mefenamic acid (Ponstel)
Analgesics (narcotic)
  Methadone hydrochloride (Adanon, Althose syrup, Dolophine)
  Morphine (trace)
  Heroin
Anesthetics
  Chloroform
  Cyclopropane
  Ether
Antibiotics and chemotherapeutics
  Chloramphenicol (Chloromycetin)
  Cycloserine (Seromycin)
  Erythromycin

Flagyl
Furodantin
Isoniazid (more than twenty trade names)
Mandelic acid
Neomycin sulfate (Mycifradin, Neobiotic)
Nitrofurantoins
Novobiocin (Albamycin, Cathomycin)
Para-aminosalicylic acid and salts (numerous trade names)
Penicillin
  G
  Benzyl
  Streptomycin
Sulfonamides (breast concentration may exceed maternal plasma level; this represents a small oral dose for infant)
  sulfamethoxazole (Gantanol)
  sulfadimethoxine (Madribon)
Tetracyclines
Antihistaminics (most pass into milk)
  Brompheniramine (Dimetane)
  Diphenhydramine hydrochloride (Benadryl)
  Methdilazine (Tacaryl)
Atropine
Barbiturates
  Amobarbital (Amytal)
  Methohexital (Brevital)
  Phenobarbital (Luminal)
  Secobarbital (Seconal)
  Thiopental (Pentothal)
Bishydroxycoumarin (Dicoumarin, Dicourmarol, Dicumarol, Melitoxin)
Bromides
Caffeine

Cortisone
Cyclophosphamide (Cytoxan)
DDT (Chloropenothane)
Diphenylhydantoin (Dilantin)
Ephedrine and pseudoephedrine
Ergot
Estrogens
Ethyl biscoumacetate (Tromexan)
Hexachlorobenzene
Imipramine hydrochloride (Tofranil)
Iodides including[1311]
Iopanoic acid (Telepaque)
Isoniazid
Laxatives and cathartics
  Aloin
  Calomel (mild mercurous chloride)
  Cascara
  Danthron (Dionone, Dorbane, Istizin)
  Rhubarb (said either not to pass or, conversely, to purge infant)
  Levopropoxyphene (Novrad)
  Mandelic acid
  Mephenoxalone (Trepidone)
  Methimazole (Tapazole)
  Methocarbamol (Robaxin)
Metals, salts, minerals
  Arsenic
  Calcium
  Chloride
  Copper
  Iodides
  Lead
  Magnesium
  Mercurous chloride (see calomel)
  Mercury
  Phosphate
  Potassium
  Sodium

Sulfur
Nicotine
Papaverine
Phenylbutazone (Butazolidin)
Phenytoin (Diphenylhydantoin, Dilantin)
Propylthiouracil
Pseudoephedrine (Sudafed)
Pyrimethamine (Daraprim)
Quinidine
Quinine
Reserpine (many trade names and preparations)
Salicylates
Scopolamine (Hyoscine)
Sedatives
  Barbiturates
  Bromides
  Chloral hydrate
  Ethinamate (Valmid)
  Sodium chloride
Thiazides
Thiouracil
Thyroid
Tolbutamide
Tranquilizers
  Chlorpromazine (Thorazine)
  Hydroxyzine (Atarax, Vistaril)
  Phenaglycodol (Ultran)
  Reserpine (many trade names)
  Trifluoperazine (Stelazine)
Vitamins
  A, B₁, B₁₂, C, D, E, K
  Folic acid
  Niacin
  Pantothenic acid
  Riboflavin
  Thiamine

Synonyms and combinations may be found in Wilson, Charles O. and Jones, Tony E., eds, *American Drug Index* (Lippincott Co. 1967). Concentrations may be found in Knowles, J.A., "Excretion of Drugs in Milk—A Review," *Journal of Pediatrics*, 66:1068, 1965.

*From J. M. Arena, "Contamination of the Ideal Food." © *Nutrition Today,* 1970.

More stringent laws should be passed regarding false claims and false accusations in advertising, and the present laws should be enforced. If we allow DDT to be banned on the basis of such propaganda as the insignificant "contamination" of mother's milk, we will most surely be burning a vital bridge—the bridge between life and death for millions of human beings who live in constant danger of insect-borne diseases.

The
Cancer
Question

5 It would be as erroneous to claim that DDT is a *cure* for a given disease as it has been for the opponents of DDT to claim that it is the *cause* of a given disease. There is no intent within this chapter to claim that DDT *cures* anything, but some interesting work, using DDT as a "medication," has been done with animals and human beings. Such cases are set forth here as incidents of interest which merit further investigation.

Toxicity is difficult to define. There is a saying among toxicologists that the subject can be easily learned in two lessons —each five years long. Nevertheless, a thumbnail study of toxicity and how it is determined must be explored before one can cite examples of any "medicinal" use of DDT. Dr. Francis A. Gunther says, "To me, toxicity means that at some dosage a given substance produces an unfavorable biological effect."

Virtually everything has a toxicity level. This does not mean that under normal circumstances a given substance necessarily causes death; however it could. The LD_{50} is a measure of *fatal* toxicity, which is determined experimentally with test animals, using an arbitrary but standard scale. The LD_{50} means that under a given set of circumstances in experimentation, with a given amount of the material being tested, 50 percent of the test animals die. LD_{50} equals the lethal dose for 50 percent of the test animals. We do not know the LD_{50} in man for most chemicals, and we can only infer it by means of experimentation with test animals.

Dr. Gunther said, "There are four basic types of toxicity: acute, chronic, dermal and inhalation. Acute toxity is a single dose (oral or injected) that in 50 percent of a theoretically large enough number of test animals will quickly kill or cause an eventually fatal biological lesion. Chronic toxicity is the amount which, given continuously over a long period of time, will produce recognizable damage. It does not have to produce death. Dermal toxicity is a response from absorption through the skin after contact of the substance with the skin. It does not have to have a lethal action. An example of dermal toxicity is the undesirable skin response to poison oak or poison ivy. The fourth type is known as inhalation toxicity. This is an unfavorable response caused by absorption through any part of the respiratory system."

Because toxicology is an extremely complex field of science, it is impossible to discuss it in full detail in one chapter of a book. The above capsulization by Dr. Gunther was prepared specifically for this book to provide a *general* background.

Comparative LD_{50}s have been established for DDT, nicotine, aspirin and many other chemicals (including all pesticide chemicals) based upon work with laboratory animals—usually several species of animals. The animals most nearly akin to man in their response to toxic substances are the higher primates—gorillas, chimpanzees, and orangutans. Because of the extreme expense and space required for these primates, together with problems of longevity and difficulty in obtaining them and reproducing them in captivity, the rat is generally used for experimentation. Rats are easy and inexpensive to raise, they reproduce readily, and they are available in standard strains. There is frequently a wide test response variation between species of animals such as rats, guinea pigs, dogs, monkeys and chickens, and sometimes between males and females within the same species. By studying the responses of various test animals the scientist is able to make a conservative estimate as to what a lethal dose might be to man.

Dr. Gunther explains, "The *lower* the amount of a given chemical needed to produce an LD_{50} in an animal, the *more* toxic it is. Four examples of estimated LD_{50}s for man, based on experimentation with rats, are these: DDT—10 grams; caffeine (in coffee)—14 grams; nicotine (in tobacco)—4 grams; theobromine (in tea and cocoa)—14 grams. This means that 14 grams of caffeine, if taken at a single time, would kill the average 160-pound man. It would take not more than four grams of nicotine to kill him."

By giving a receptive animal an abnormal amount of any given substance by methods of administration never encountered outside of the laboratory, and to the exclusion of a regular well-balanced diet, a scientist is often able to produce in that animal a wide range of adverse effects. These effects would not occur under normal environmental conditions. Such methods have frequently been used to mislead and alarm the public into fearing minute traces of pesticides and food additives. Caution rather than alarm is in order. It is easier to *cause* ill effects than to cure, alleviate, or remove them. No experimentation with test animals is meaningful unless there is a control group of untreated test animals for comparision. Human beings will not necessarily respond to a test substance in the same manner as the test animals. Obviously it is necessary to use test animals for experimentation; however, regardless of the method of administration, the dosage, or the material being tested for curative or preventive purposes, the time inevitably comes when the process or material is used for the first time on a human being.

Dr. Edward R. Laws, aided by Dr. Richard T. Rappolt, Sr., conducted the studies at the Montrose Chemical plant that were discussed in Chapter Three. In an interview, Dr. Rappolt explained that the primary purpose of going to Montrose was to learn more about the effects of DDT on people. He remarked that the subject of cancer came up by accident when an employee asked the doctors what they were doing. The employee listened to the doctors' brief explanation and then, according to Dr. Rappolt, "He casually commented, 'Well, I can tell you one thing. Nobody around here ever dies of cancer.' " The doctors were skeptical about such a casual statement, but they *did* pursue it. After carefully checking the files of every Montrose employee in the plant's history, which at that time constituted 1,300 man years of work, the doctors found that the records verified the workman's statement. There had, in fact, been no deaths from cancer. Dr. Laws said, "It is curious that no cases of cancer had developed in some 1,300 man years of exposure, a statistically improbable event."

In contrast to the Montrose study and other incidents to be cited in this chapter, Rachel Carson stated on page 225 of *Silent Spring;* "Dr. Hueper now gives DDT the definite rating of a 'chemical carcinogen.' " She cited as her reference an article written by W. C. Hueper, and W. W. Payne, "Observations on the Occurrence of Hepatomas in Rainbow Trout."

At the time of the Montrose study, Dr. Laws was employed by the United States Department of Public Health, and Dr. Rappolt was employed by the California State Department of Public Health. Dr. Laws later accepted a residency in neurosurgery at Johns Hopkins Hospital. At this writing he is Resident-in-Chief, Neurological Surgery, The Johns Hopkins Hospital, Baltimore, Maryland. At this hospital, Dr. Laws began extensive experimentation on mice to test the hypothesis that DDT has an anti-carcinogenic effect. His findings have been published in the *Archives of Environmental Health,* Fall, 1971.

For this research, Dr. Laws used a strain of mice which had been tested over a ten-year period and found to be 100 percent susceptible to the development of terminal malignant brain tumors. The tumors were transmitted by subcutaneous injection.

Over a period of sixteen months, Dr. Laws te*
He divided the mice into two groups, one for ex~~
and the other to be used as a control. The mice were five we~~
old before experimentation was started. In order to eliminate
hormonal differences between the male and female mice, only
male mice were used. The control mice were placed on a diet
of SLC (Standard Laboratory Chow). The experimental mice
were placed on the same diet to which 33.3 mg/kg (ppm) of
DDT had been added. This amount is approximately 10,000
times the average daily human consumption of DDT. Dr. Laws
kept both groups on this diet for one week and then injected
them with the tumor-producing suspension. Nearly all the experimental animals developed tremors and hyperexcitability
lasting from three to seven days.

A summary of Dr. Laws' experiment showed that 100 percent
of the mice in the control group died of brain cancer. In the
experimental group 92.1 percent died of brain cancer. The longevity of the animals in the control group was 40.1 $\pm$ 4.1 days,
while the longevity of the mice which did die of cancer in the
experimental DDT group was 56.5 $\pm$ 5.8 days. The fact that
7.9 percent of the animals on the DDT diet survived may or
may not be significant. We *do* know that prior to this work and,
over a ten-year period, 100 percent of the injected mice died of
cancer. Further experimentation is certainly needed in this
area.

Dr. Gunther has said, "The livers of warm-blooded animals
have never adequately adapted to cope with any organic compound containing carbon-chlorine bonds except in very minute
amounts over a long period of time and in an extremely ineffi-
cient manner. (An exhaustive search of the world's chemical
literature has revealed that extremely few compounds which
contain carbon-chlorine bonds exist in nature from plants, bacteria or molds. One of these is the well known antibiotic chloromycetin). Each DDT molecule has five carbon-chlorine bonds;
however, this figure is of no significance to human beings be-

cause of the *minute* amounts of DDT in our daily diets. It has been established that DDT is an enzyme inducer in the liver. This type of enzymatic action can cause poisons and other foreign materials to be broken down and/or eliminated." Is it possible that uncommonly *high* levels of carbon-chlorine bonds in men and animals inhibit cancer cell growth?

An article entitled "New Cancer Drugs Studied—A Few Found Promising" appeared in the February 1971 issue of *Infectious Diseases*. The staff writer said: "Help from an unlikely source is present in the form of o, p'-DDD [action], a derivative of the insecticides DDT and DDD. This compound serves one function, but it does it uniquely; it is organ-specific, acting only on the adrenals. It is the only known drug that is effective against adrenal carcinoma, and has a clinical response rate of 40 percent.

"The mechanism of o,p'-DDD [action] is unknown, but it is thought to suppress steroid hormone production through inhibition of enzymes. Studies of blood and urine have shown a rapid reduction in the levels of adrenocorticosteroids and their metabolites. Side effects include anorexia, nausea, somnolence, and dermatitis."

Ironically, on February 16, 1971, James M. Staples, reporter for the *Evening News*, Newark, New Jersey, wrote: "A pioneer scientific foe of the pesticide DDT has taken issue with a report from the World Health Organization of the United Nations which favors continued use of the controversial chemical to combat malaria mosquitoes in under-developed countries.

"Dr. Charles F. Wurster, assistant [associate] professor of Biological Science at the State University of New York, at Stony Brook, L. I., and an organizer and leader of the Environmental Defense Fund, charged the U. N. organization with perpetuating the use of a known cause of cancer.

" 'They are busy adding a carcinogen (cancer inducer) to the tissues of all humans on earth and they call themselves the World Health Organization,' said Wurster.

" 'They are adding to human tissue a material shown to be

carcinogenic in several species of laboratory animals, which indicates a very high probability that it is also a carcinogenic hazard to humans,' Wurster said."

Honorato S. Echavez, a native of the Philippines, and a former officer in the United States Army, was in charge of the engineering phase of the Malaria Eradication Project in Burma for WHO between 1961 and 1963. Mr. Echavez earned two degrees in engineering from the University of Kansas, and an M.S. in Sanitary Engineering from Johns Hopkins University. He has tried in vain since 1961 to encourage research regarding an unusual medical phenomenon he witnessed in Burma.

The following is part of a personal letter from Mr. Echavez, written on February 20, 1970:

A Burmese tried to commit suicide by drinking a glass of DDT solution. The reason for this suicide attempt was that he was suffering from leprosy which was quite common in that region. Instead of dying from DDT poisoning, he not only survived but recovered from leprosy after a two weeks period. According to Dr. Way, who was then Chief of the Malaria Eradication Project in that Sector of Burma, other lepers did the same thing and were cured of leprosy.

I asked Dr. Way and Dr. P. P. Baroa, who confirmed this report, why they didn't do further research or write an article covering this medical phenomenon. They said that they didn't want to get involved because DDT is believed to have deleterious effects on both the liver and kidneys. (Dr. P. P. Baroa is the Chief of the Malaria Eradication Project in the Mandalay Sector).

This incident described above happened in 1961 in the sector in which Taunggy is the Headquarters. Taunggy is some 130 air miles southwest of Mandalay, or some 300 miles NNE from Rangoon, the capital.

Very sincerely,
H. S. Echavez

It is not known what constituted the solvent for the DDT solution. Just as in the suicide cases mentioned in Chapter Three, where DDT was *not* listed as the cause of death, DDT in this case cannot be claimed as the cure; *however,* since there is no record of anyone having suffered any deleterious effects from DDT on either the kidneys or the liver, why have these leprosy incidents been neglected for over twelve years? Why has Mr. Echavez's appeal for further study in this area been ignored?

Dr. Laws' colleague in the Montrose Study, Dr. Richard T. Rappolt, Sr., received extensive news coverage when he used five grams of DDT in warm peanut oil to treat three women who were admitted to Park Emergency Hospital in San Francisco in 1969 for barbiturate overdoses.

The *San Francisco Examiner,* July 17, 1970, carried the following banner headline on page one: "Bizarre Medical Experiment—3 S. F. Patients Given DDT." Robert Gillette, reporter for the *Examiner,* wrote, "In a brief but potentially disastrous medical experiment a city health department doctor has given massive internal doses of the pesticide DDT to patients in San Francisco hospitals. The physician, Dr. Richard T. Rappolt, Sr., induced three women last year to swallow DDT to test whether the insecticide would counteract an overdose of sleeping pills."

The reason cited by Gillette was speculation. Within his article he erroneously claimed that the five-gram dosage which was given technically carried a 50 percent risk of killing a normal 110-pound adult. The estimated LD_{50} for DDT is 10 grams for a 160-pound man based upon experiments with rats; therefore, 10 grams involves a 50 percent risk, but 5 grams involves no risk or only a very slight risk. After stating that Dr. Rappolt *induced* the three women to swallow DDT, Gillette said, "All three women, age 18 to 39, were still stupefied by sleeping pills when Dr. Rappolt gave each the equivalent of one teaspoon of pure DDT mixed in a small amount of warm peanut oil."

Later in the article Gillette quoted Dr. Rappolt as saying he had consent from one patient and from relatives of the two others. In the next paragraph, as a gratuitous slap at Dr. Rappolt, Gillette said, "But many researchers would question whether consent from a drugged patient or from flustered relatives in an emergency ward is really valid."

In his article Gillette admitted that it has been common knowledge for twenty years that people have taken and survived massive doses of DDT. In all three cases treated by Dr. Rappolt, the patients were discharged and able to walk out of the hospital the next day. Dr. Rappolt had taken 5 grams of DDT himself before he ever administered this amount to a patient. He said that he underwent several hours of nausea, several nights of insomnia and six to seven days of excitability and nervous tension.

Dr. Rappolt subsequently prepared a paper which he planned to present before the Seventh Inter-American Conference on Toxicology and Occupational Medicine. A newsman obtained a copy of this paper and headlines appeared from San Francisco to Miami. Although Dr. Rappolt treated the women in January and February 1969, the story was not printed until July 1970. Dr. Francis J. Curry, Director of Public Health, City and County of San Francisco, subsequently fired Dr. Rappolt *nearly a year and a half* after the patients' cure. It is Dr. Rappolt's contention that since DDT is an enzyme inducer, the action of DDT counteracts and causes detoxification of barbiturates by stimulating the liver to produce more enzymes.

In a follow-up story, the *Miami Herald* ran this headline: "Experiment with DDT Dosages Ruinous to Dr.'s Career." Dr. Rappolt's career has been, in fact, far from ruined. He has a successful practice in San Francisco, is executive editor of *Clinical Toxicology,* and is chairman of the Drug Abuse Committee of the American Academy of Clinical Toxicology.

Dr. Rappolt has never attempted to conceal his treatment. Not only does he possess photostatic copies of the charts for each patient clearly stating that he had administered 5 grams of DDT, but also he openly discussed these cases at the State of Washington DDT Hearings in October, 1969. He openly presented the same information to the Seventh Inter-American Conference on Toxicology and Occupational Medicine. Was further study on the use of DDT in barbiturate poisoning abandoned because of anti-DDT publicity? Would Dr. Rappolt have been fired and subjected to such adverse publicity if his treatment had involved a less controversial pesticide? He was criticized for not obtaining the permission of his superiors. Would there have been time? Would they have allowed it in the light of the anti-DDT publicity? The only things known for sure are that all three patients survived and that Dr. Rappolt is now a physician in good standing on the staffs of three San Francisco hospitals.

In all cases cited in this chapter, one must leave the final evaluations to the medical profession; however, the profession should not abandon or reject research on the possible beneficial aspects of *any* chemical simply because that chemical is unpopular or controversial. *Res ipsa loquitur.* Let the facts speak for themselves.

Is the Spring Silent?

6

The alleged decline or disappearance of several species of birds, some symbolic, some common, and some rarely heard of, at least by their scientific names, has become another tool in the hands of the environmentalists in their relentless attempts to ban DDT. It is impossible to prove the absence of any biological effect whatsoever. It is also impossible to prove the total absence of any chemical in any system; hence the charges that DDT *could* have caused, *might* have caused, *possibly* caused, or *probably* caused thin eggshells are impossible to prove or disprove.

Numerous researchers have attempted to prove that DDT or one of its metabolites, particularly DDE, causes thin eggshells. The results of such experiments are confusing to the layman and, in some instances, have also confused the researchers.

Many factors are known to contribute to thin eggshells or to lack of hatchability. Diet, weather, noise, excitement, and a wide range of chemicals—including the omnipresent PCBs, lead, mercury and dieldrin—are just a few of these factors. The anti-DDT activists insist that DDT and DDE cause thin eggshells and that their use has resulted in the decline of certain species of birds. Rather than explore their unrealistic, inconclusive laboratory experiments, it is more germane to look instead at some of the birds alleged to be "endangered."

Non-Predatory Birds

In *Silent Spring,* Rachel Carson devoted an entire chapter, ominously titled "And No Birds Sing," to the alleged effects of pesticides on bird life. The basis for this science fiction title is vague. On page 104 she quoted *an observer* who reported that she had seen "no land birds at all for long distances" while she was driving through southern Mississippi. On pages 104-105 Miss Carson wrote: "One story might serve as the tragic symbol of the fate of the birds—a fate that has already overtaken some species and that threatens all. It is the story of the robin, the bird known to everyone. To millions of Americans, the season's first robin means that the grip of winter is broken. Its coming is an event reported in the newspapers and told eagerly at the breakfast table. And as the number of migrants grows and the first mists of green appear in the woodlands, thousands of people listen for the first dawn chorus of the robins throbbing in the early morning light. But now all is changed, and not even the return of the birds may be taken for granted."

The Audubon Society reveres Miss Carson and has established a fund in her memory. Miss Carson's book was published in 1962. A comparison of the Audubon Society's Christmas Bird Count of 1941 (prior to DDT) with their 1960 Bird Count indicates that there was a 1137 percent increase in robins per observer. This information was readily available to Miss Carson *before* she wrote *Silent Spring.* Also included on her "threatened species" list is the swallow. On page 111 she said, "The loss of food has also struck hard at the swallows that cruise the skies, straining out the aerial insects as herring strain the plankton of the sea." The above, though lyrical, is non-factual, since the Audubon Society reported a 341 percent increase of swallows per observer between 1941 and 1960.

Predatory Birds

The Hawk Mountain Sanctuary Association has been counting predatory birds on the Appalachian Flyway in Pennsylvania since 1935, with the exception of the war years, 1943, 1944, and 1945. Many predatory birds including eagles are tabulated by the Association under the general category of "hawks."

Fluctuation in hawk numbers is attributable to population cycles, and to such variable factors as weather and available food supply. Several birds in the "hawk" category have been used by the opponets of DDT as "proof" that DDT should be banned because it allegedly contaminates the food supply of predatory birds and causes thin-shelled eggs. The data for the five "hawks" shown on the chart (page 72) illustrate the fluctuation in their populations over a thirty-year period. The figures in the "Total" column include *all* hawks sighted at the sanctuary, not just the five examples shown. It is curious to note the unusual rise and fall of the peregrine falcon count.

THE OSPREY   The osprey is a large bird of prey, commonly called the fish hawk. These birds live near either fresh or salt water in nearly every temperate and tropical country of the world. They are found in the higher latitudes and altitudes during the summer, but because the frozen lakes and streams make food unavailable during the winter, they, like many other birds, migrate to warmer areas in the winter. Standing approximately two feet tall with a wingspan of nearly six feet, the osprey feeds on the fish which swim close to the surface of the water. It is amusing to watch one of these unusual birds as it enters the water with a great splash. Despite its explosive entrance, it usually emerges with a fish.

"The osprey is in trouble," lamented Roger Tory Peterson in *National Geographic Magazine,* July, 1969. "We know that DDT and other chlorinated hydrocarbon pesticides sprayed widely over farms and orchards persist in the environment for a long time. Insects poisoned by these chemicals are consumed

PART OF HAWK MOUNTAIN REPORT

YEAR	RED-TAIL HAWK	GOLDEN EAGLE	BALD EAGLE	OSPREY	PEREGRINE FALCON	TOTAL* ALL HAWKS
1935 —	4,024	66	67	169	14	15,765
1940 —	4,725	72	38	91	25	11,236
1946** +	2,358	69	42	191	26	9,291
1950 +	2,669	68	142	323	35	14,916
1955 ++	3,802	57	89	359	35	20,191
1960 ++	2,317	38	37	303	26	18,893
1965 ++	3,305	34	43	444	14	17,986
1970 /	3,632	25	28	600	27	24,000

LEGEND: — = before DDT
+ = small amounts of DDT used in America
++ = highest period of DDT use in America
/ = minimal amounts of DDT used in America

*TOTAL column includes all hawks counted, not just five examples shown. Examples chosen because of publicity concerning them.
**No records are available for the years of World War II.

by fingerlings; these are eaten by larger fish which in turn are caught by the osprey. The concentrated chemical poisons are then presumably transferred to the osprey's own tissues. Recent studies at the University of Wisconsin show a definite relationship between the presence of DDE—a derivative of DDT—in the eggs of certain fish-eating birds and the thickness of the eggshells. Osprey eggshells examined in New Jersey weighed about 25 percent less than those collected before the use of DDT."

In the concluding sentence of the article Mr. Peterson said, "One year soon, I fear I shall go down to Great Island, at the usual time of their arrival and there will be no osprey—not one."

In 1967 there were 403 osprey counted at Hawk Mountain. At the time of Mr. Peterson's 1969 article the count was 530 and 600 were counted in 1970—a new record high.

The bald eagle, incidentally, competes with the osprey for food. On many fishing trips in Montana I have seen an osprey perched on a snag or hovering over a stream, intently watching the water. Suddenly the bird would drop into the water with an enormous splash and emerge with a fish. Before it could get back to its perch, a bald eagle would swoop down, snatch the fish from the osprey, and without a pause fly out of sight.

Chemical pollution in and along Long Island Sound has caused a decline in the osprey population *in that area.* Nevertheless, the osprey is far from a vanishing species. According to the latest available report from Hawk Mountain, osprey migration increased by approximately 86 percent between 1950 and 1970. In the Lake Coeur d'Alene, Idaho, watershed area, the number of active osprey nests increased 925 percent during the same twenty-year period.

GULLS AND TERNS   In May, 1971, two oil tankers collided in the fog under San Francisco's Golden Gate Bridge, spilling an estimated 800,000 gallons of fuel oil into the San Francisco

Bay. This collision was responsible for killing an estimated 3,000 birds, including grebes, scooters, loons, ducks, gulls, and murres. An additional 4,000 oiled birds were "rescued" and taken to the San Francisco Zoo and to a special center in nearby Richmond, California. A volunteer army of bird-lovers rushed to the scene in a desperate attempt to save these birds. As a result of their efforts, 305 birds, including gulls, were saved at an estimated cost of 400 dollars per bird. This information was provided by Irston R. Barnes, past chairman, Audubon Naturalists' Society. Mr. Barnes estimated that 120,000 dollars were spent on this effort (not including the donated man hours). UNICEF figures show that in underdeveloped countries, 120,-000 dollars would provide an adequate diet for twelve thousand children for one month or one thousand children for one year.

Just two weeks prior to the San Francisco oil spill, the Audubon Society was at work near Cape Cod setting out poison to *kill* seagulls. A *New York Times* article, April 18, 1971, headlined "Audubon Society To Poison Gulls," reported how the Audubon Society, with the permission of the Massachusetts Department of Natural Resources, poisoned many thousands of gulls despite the fact that these birds are on the Massachusetts "protected list" of sea birds. They baited the gulls with chunks of bread covered with a fatal mixture of margarine and a drug called No. 1339. "It's kind of like weeding a garden," explained Dr. William H. Drury of the Massachusetts Audubon Society. "The gulls have taken over a traditional nesting place of terns and the Society will try to preserve the terns by eliminating the gulls from one island."

The results of a study of the tern population on Great Gull Island (Long Island Sound, New York), by Helen Hays and Dr. Robert Risebrough (a science advisor to the Environmental Defense Fund), appeared in *Natural History Magazine,* November, 1971. After an extensive discussion of the local decline in the tern population and of abnormalities in tern chicks, the authors wrote, "Added to this, in 1971 a new abnormality was observed in the colony for the first time: thin-shelled eggs not sufficiently strong to support the weight of the incubating parents." DDT was first used in the United States in the mid-1940s. Why would terns suddenly lay thin-shelled eggs for the *first* time, in 1971, if it were a result of DDT? Yet the authors state, "Researchers in this area believe that shell-thinning is caused primarily by the DDT compound, DDE." They said also that, "DDE has been shown to produce shell-thinning under experimental conditions in mallard ducks."

This mallard duck study, which was conducted by Robert Heath, must be discussed here because it was the vortex of the eggshell propaganda. Human beings ingest up to 0.03 mg of DDT per day. This DDT is broken down *within our bodies* into an also infinitesimal amount of DDE. Heath fed some pens of experimental ducks 2.5 ppm (mg/kg) of DDT, and other pens 10 ppm of DDT for two years (1965–1966). These ducks showed a great *increase* in the number of ducklings produced, despite the fact that the amount of DDT in their daily diet was 100 to 400 times greater than the daily intake of a 160-pound human being.

DDE is a metabolite of DDT and is *never* used as a pesticide. In the environment, this product results mainly from the natural metabolic breakdown within birds, fish, and mammals and in soils.

During 1967 and 1968 Heath fed some pens of ducks 10 ppm, and others 40 ppm of synthetic DDE mixed with their food and then announced that his experiment *proved* that DDE caused thin eggshells. If ducks eat DDT it *is* broken down in their bodies to DDE. Heath's feeding experiment with DDE showed only that DDE, at astronomically high levels (400 to 1600 times the human norm), inhibited reproduction.

From Heath's 1967–1968 experiment with DDE, his colleagues concluded that the use of DDT as an insecticide resulted in the buildup of DDE in birds, thus causing reproductive failures. The fallacy of this theory is immediately evident because the ducks in his first experiment which were fed on prodigious amounts of DDT actually produced 81 percent *more* ducklings than the control group which received no DDT or DDE in their diet. There has been little, if any, publicity given to the results of the first experiment in which DDT was used, while widespread publicity has been given to the second.

Heath used DDE because predatory birds are at the top of the food chain and ingest DDE, which is present in their prey. Nevertheless, a bird would have to consume a fish of "Moby Dick" proportions daily, even to approach the level of DDE which Heath's ducks consumed.

THE PEREGRINE FALCON   In an advertisement in the October, 1970 issue of *Audubon Magazine,* the National Audubon Society and the Environmental Defense Fund joined forces to capitalize on the alleged decline of the peregrine falcon. At the top of the page is a picture of the head of a peregrine falcon. Next to the picture, in large boldface type, is the question, "Have you seen a peregrine falcon lately?" In the full-page advertisement the sponsors reiterate the charges that DDT causes eggshell thinning and urge that this pesticide be banned. Also included is a large coupon for the reader to mail with his check payable to the EDF for memberships ranging from five dollars for a student to five thousand dollars for a founding member. *Have* you seen a peregrine falcon lately? Have you *ever* seen a peregrine falcon? Would you *recognize* a peregrine falcon?

J. J. Hickey's book, *Peregrine Falcon Populations—Their Biology and Decline,* published in 1968, is by the very nature of its title misleading, because it implies an unwarranted conclusion, based on inconclusive "evidence." In the final chapters Hickey postulates a world-wide decline of the species from evidence collected entirely within the United States and one small part of Europe. This conclusion is drawn despite the fact that it was certainly known and even pointed out in the introductory chapters of his book that the peregrine has its highest breeding densities and major populations in those areas of the world in which human populations are low. The major "evidence of declines" is taken from observations of specific regions where human population density is very high and where, in addition, massive and wide-spread ecological damage has occurred.

Frank L. Beebe, a Canadian scientist who has been studying the peregrine for twenty years, has commented: "The myth of the vanishing peregrine has been built around a noticeable reduction in the number of nesting pairs of this falcon in two regions of the world. The decline in these regions is well documented and not questioned. The reason for the decline in these areas is well known. To say that the peregrine falcon as a species continent-wide or world-wide, is declining at all, let alone declining to the point of being endangered is questionable, and insofar as its population in the Canadian Arctic and the Canadian Pacific Coast, is denied."

In their article, published in *The Condor,* 1968, Anderson and Berger note, "Peregrine falcons in Alaska, which are reproducing successfully, have DDT levels twice as high as those in the Eastern United States, which show a decline in number. There are obvious causes other than DDT for the decline of falcons; in common with other large predators they depend on a wilderness environment for their habitat."

It is difficult to believe that any species that has the peregrine falcon's biological resilience and ability to breed from the Arctic and the sub-Antarctic to the equator, from sea level to elevations of 14,000 feet, and from rain forest humidity to Saharan aridity, is a convincing candidate for immediate or early extinction. The fact that the peregrine falcon population has diminished in *areas of heavy human population* does not make it a candidate for "extinction."

Regardless of how much the peregrine falcon is extolled and glamorized by those seeking a ban on DDT, the unpleasant ecological fact remains that the peregrine falcon, throughout approximately 90 percent of its world-wide range, is a weed species, commonly called the duck hawk. It is an aggressive, rapacious invader of the ecosystem, in which it competes with other more specialized and regionally confined birds. This bird kills far more than it can possibly eat, thus robbing many other predatory birds of their food supply.

THE EAGLE   Rachel Carson wrote, "Like the robin, another American bird seems to be on the verge of extinction. This is the national symbol, the eagle. Its populations have dwindled drastically within the past decade. The facts indicate that something is at work in the eagle's environment which has virtually destroyed its ability to reproduce. What this may be is not yet definitely known, but there is some evidence that insecticides are responsible."

The eagle population in the United States has shown a marked decline for many years. This decline was first noted in 1917, twenty-six years before DDT was introduced in this country. In 1921 an article in *Ecology* warned that the bald eagle was threatened with extinction. Since DDT had not yet been used, it could not have been incriminated. Man with his guns, his traps, his coyote poisons and his unrestricted shooting has drastically lowered the number of both golden and bald eagles. In 1917 Alaska placed a bounty on the bald eagle and between 1917 and 1952 paid bounties on 115,000 eagles. In 1921 W.G. Van Name wrote, "The bald eagle is now fast becoming a rare bird in the United States, probably much rarer than most people or ornithologists are aware." Nevertheless, the Audubon Christmas Counts showed a 25 percent increase in the contiguous forty-eight states between 1941 and 1960, indicating that these birds were still reproducing, and that, within that twenty-year period, they had held their own and even shown an increase.

The eagle is not seen in areas where it was once seen, because of housing developments, logging, and the resultant lack of food. The decline of the bald eagle, though not a recent development, as our environmental propagandists would have us believe, has become a *cause célèbre* because the eagle is our national symbol.

The term "bald eagle" is a misnomer. This eagle does, in fact, have feathers on its head. It is the vulture that is bald. When these eagles are young, they are gray or smoky brown, with a few white spots. This makes it very difficult to differentiate between a young bald eagle and the golden eagle which also has white spots, even though the golden eagle is a deeper brown. These birds fly to heights exceeding 1,000 feet. Positive identification is virtually impossible when they are high aloft.

When the bald eagle is three to four years old, the head, neck and tail feathers turn white. The bill, feet, and bare leg parts are yellow. By contrast the golden eagle has feathers down to the talons. The mature bald eagle is approximately three feet tall with a seven- to eight-foot wingspan and a weight of eight to twelve pounds. The eagle's nests (eyries) are enormous. I have seen one such eyrie in Glacier National Park that was at least six to seven feet in depth and approximately the same width at the top.

On one visit to Glacier National Park, I listened to a ranger at a campfire talk. He told the group that the American Bald Eagle was "disappearing into a vanishing species." The following night a different ranger at the same campsite told us that more than 280 bald eagles had been counted migrating up the Flathead River near park headquarters *in a single day.* I wrote to George Hartzog, Director of the National Park Service in Washington, D. C., to ask for a statement regarding these two conflicting reports. My letter was answered by Robert Linn, Ph.D., Chief Scientist, National Park Service. He said, "The two separate pieces of information in Glacier concerning the American Bald Eagle were both probably correct. The bird may well be declining and it is not unusual to see fairly large groups of them together during fish migration in the rivers along their migration route." Dr. Linn also provided the following information: (All conditional clauses are italicized to stress the conjectural nature of this statement.)

The Southern Bald Eagle that ranges from Florida to Canada on the East Coast of the United States is on the Endangered Species List. The Northern Bald and Golden Eagles of the West that range up into Alaska are protected by Federal Law. The evidence *seems* to indicate that the numbers of American Bald Eagles are declining and one of the reasons cited is the failure of egg hatch, *presumably* due to interference with the calcium metabolism of the bird which results in thin eggs and hatch failures. The evidence for the decline of the species is *presumably* found in the numbers of abandoned nests and the *apparent* decrease in numbers of young birds. As you may have been reading in the papers, numbers of dead eagles are being found in Wyoming and analysis of the carcasses is being made by the Bureau of Sport Fisheries and Wildlife. It is *conjectured* that the Wyoming birds are being killed by poisons distributed for coyotes and other predators.

Frank Thone called attention to the declining bald eagle population in "Nature Ramblings," *Science News Letter,* July, 1943. He noted that most eagles live in trees near fish-filled rivers and not on inaccessible cliffs. "So when the timber was cleared," Thone said, "it was inevitable that the eagle had to go. Moreover the cities grew and befouled the rivers with sewage and industrial wastes. The once teeming fish population vanished. With their main source of supplies thus taken away, it was only natural that the eagles should vanish also." This article was written several years before the first extensive use of DDT.

It is true that some eagles nest in trees, but many others definitely nest in inaccessible cliffs. Eagles do not "live" anywhere. They nest in a specific area, but they travel over vast areas in search of available food. Bald eagles are commonly called fish eagles. They often perch on snags or very large trees.

Considering the size of the bird and the size of the nest, it is obvious that the eagle would require a tree of considerable size on which to build the nest. Hence this and most other large predatory birds are not found in heavily populated areas.

In an Associated Press release of May 10, 1971, Wyoming's State Representative, John Turner, is referred to as an authority on eagles. In this article Turner is quoted as saying, "In addition to the discovery of eleven poisoned bald eagles and four golden eagles in Wyoming, 25 others were found shot in the ditch, 90 miles southwest of Casper, Wyoming." Representative Turner, who at this writing is working on his doctorate in wildlife ecology at the University of Michigan, stated, "There are considerably less than 200 mated pairs of bald eagles left in the country."

Less than three months later, on August 2, 1971, James O. Vogan, a helicopter pilot for the Buffalo Flying Service, Buffalo, Wyoming, testified before the United States Senate in regard to the slaughter of eagles. Vogan said that he had seen 560 rare eagles killed in Wyoming by shotguns from airplanes during the preceding winter months. "We had a regular haystack of these eagles when we first started bringing them in," he recalled. Vogan said that he had been told by his boss, Doyle Vaughn, that Vaughn had a permit to shoot the eagles and that he told the gunmen the same thing. Although Vogan never saw the permit, he did not question that the killings were legal. Originally, according to Vogan's testimony, Wyoming ranchers paid 25 dollars per eagle but because so many were killed the ranchers began to pay at the rate of 80 dollars per day. The pilot telephoned Wyoming Senator Gale McGee to report this slaughter after he had read about the senator's investigation earlier that year into the death by poisoning of twenty-three eagles in Wyoming.

Federal law prohibits the killing of the bald or golden eagle except under special permits, accorded when a governor certifies that they endanger livestock. According to the Associated

Press story, Senator McGee guaranteed Vogan immunity from prosecution in exchange for his day-by-day account of the eagle killings and the names of the individuals involved.

Vogan testified that three men alone had shot thirty-three eagles and fourteen coyotes one day, and thirty-four eagles and twelve coyotes the next. He said that the sheep ranchers claimed that they had to kill off the coyotes and eagles because they were eating newborn lambs. Vogan said that he had never seen a bald eagle attack a lamb but had seen a golden eagle flying off with a lamb in its talons. Nevertheless, in his view the coyote was the main threat to the lambs. Later in his testimony, Vogan said that he had instructions to stack the dead eagles in a sheep-shearing shed and that ranchers finally dug a long trench and buried the eagles with cattle carcasses.

This story clearly points to the gun as the cause for a decline in eagle population, and certainly careful and more precise legislation is needed to curb wanton slaying.

The National Park Service reported that 267 bald eagles were counted during the *week* of November 12, 1971, in Glacier National Park. The United States Department of the Interior estimates that there were about 7,000 bald eagles in Alaska in 1970. Yet the anti-DDT forces continue to claim that the bald eagle is being exterminated by DDT, and new stories and books containing this fallacious information continue to appear.

THE PELICAN   The brown pelican is a large webfooted, fish-eating sea bird common along the coasts of Southern California, Baja California, and Florida. After scanning the water from the air, the brown pelican dives into the water and uses the large pouch under its bill to scoop up small fish. Its movements on the ground are awkward, but it is a fast swimmer and a strong, graceful flyer. Although the brown pelican is indigenous to all of the areas mentioned above, the environmentalists have singled out Anacapa Island, the smallest island within the Channel Islands National Monument, as their "proving ground" for the alleged decline in the pelican population. The Channel Islands are a chain of islands extending southward from a point off Santa Barbara, California, to Santa Catalina. They include the islands of Santa Cruz, Santa Rosa, Santa Barbara, San Nicolas, San Miguel, Santa Catalina and Anacapa.

Anacapa is the most accessible to man because it is only thirteen and one-half miles off the California coast, and because it is easy to circle by boat. This small island (approximately 3,000 feet by 5,000 feet) was unknown to most Americans until just a few years ago when it became the center of controversy over the rise and fall in brown pelican populations and DDT's alleged role in this fluctuation.

In 1923, William L. Dawson, a well-known bird expert, visited Anacapa and was dismayed by the disaster caused by his visit. The pelicans rapidly left their nests when he came ashore, and the gulls promptly pecked open the pelican eggs. Because of this, Dawson urged that no one ever go ashore on Anacapa again.

In his 1966 article, "Terrestrial Vertebrates of Anacapa Island, Calfornia," *San Diego Society of Natural History,* R. C. Banks said, "There has been apparently little change in the size of the population since the earliest reports [1898]."

In *Audubon Field Notes,* 1969, Schreiber and DeLong stated that while fluctuation in numbers do occur from year to year, there had been no apparent decline in recent years along the California coast. Members of the Pacific Ocean Biological Survey Program of the Smithsonian Institution visited the islands April 5, 1968, and recorded more than two hundred adult birds which were thought to be nesting. On another visit on May 13, however, they recorded that the birds were "apparently not nesting." Could the birds have merely completed their nesting during that six-week interval? In the same article, DeLong said that on the evening of July 21, 1968, while he was aboard ship, he observed forty *young birds* on or near the nest. He said,

"This is not a complete census as not all of the area could be seen."

On January 28, 1969, the famous Santa Barbara oil spill occurred, resulting in severe damage to nearly every kind of seashore life in that area. Dr. Robert Risebrough, a biochemist at the Institute of Marine Resources, University of California, Berkeley, and a science advisor for the Environmental Defense Fund, went with three colleagues to Anacapa on March 18, 1969, to check the damage of the oil spill. In 1970, J. R. Jehl discussed Risebrough's visit in an article in *Pacific Discovery.* He said, "It was the nesting failure of pelicans on Anacapa, hard on the heels of the Santa Barbara oil slick, that demonstrated the extent to which our marine waters had become contaminated." He then added, "As soon as the condition of the eggs was appreciated, Risebrough was able to postulate that chlorinated hydrocarbon pesticides were probably responsible."

During the spring and summer of 1970, one of Dr. Risebrough's associates, Franklin Gress, supported by a National Science Foundation grant, the National Audubon Society, and the California Department of Fish and Game, made ten trips to Anacapa between February and August. Access to the island was provided by Rotoraid, Inc., a helicopter service based at Ventura Marina. In a report he prepared, Gress stated that he arrived on Anacapa on March 29, 1970, and found 127 nests, many containing eggs. During his three-day stay, the number of nesting birds gradually dwindled as nests were deserted. Later he stated that when a new colony was observed on April 20, only sixty-five nests were counted, with approximately 120 pelicans on the nesting areas. By April 22, two days after Gress had arrived by helicopter, this new colony was suddenly deserted also.

Gress made ten trips to the island, spending a total of thirty-six days there. The abstract of his lengthy reports reads:

Recent evidence of reproductive failures of the brown pelican in California prompted a study of the Anacapa population in 1970. From February to August reproductive success and breeding biology were studied. There were 552 nesting attempts in two colonies established on Anacapa in 1970, with but one young produced. Reproductive failure was attributed to thin eggshells which collapsed during incubation. Aberrant behavior associated with reproduction was also observed. In a survey of all traditional breeding sites in California no other nesting was reported. Thus, in the State of California, only one young pelican was hatched in 1970. Double crested cormorants nesting on Anacapa suffered the same fate as the pelicans. In at least 50 nesting attempts, only one nest, producing three cormorants, was successful.

There has been no documented correlation between DDT in our coastal waters and the sudden catastrophe that befell the pelican eggshells on Anacapa in March, 1969. There *was* a close correlation in time between the Santa Barbara oil spill and thin eggshells that developed just two months later *for the first time in recorded history.* Excitement is a common cause of thin eggshells. The greatest excitement the Anacapa pelicans have had in many years was the repeated helicopter landings during the investigation of the oil spill in March, 1969, and during Gress's investigations in the spring and summer of 1970. His statement that only one young pelican was produced in 1970 misled the whole nation to believe that the brown pelican was doomed in California.

Gress did his pelican study on Anacapa Island and based his conclusions upon that study, yet according to superintendent Don Robinson, Channel Islands National Monument, "Anacapa was *never* the main breeding site of the pelican."

Recalling that the presence of Mr. Dawson in 1923 had caused considerable disturbance to the pelicans, Dr. J. Gordon Edwards, noted that many helicopter landings and the recurring presence of a human being tramping about on the island for a total of thirty-six days would obviously have an adverse effect on the thickness of eggshells and the nesting success of these shy birds. Dr. Edwards wrote several letters in an attempt to get support for his proposal to ban helicopter landings on Anacapa. He reports that his letter to the California Department of Fish and Game was fruitless and that neither the Audubon Society nor the Sierra Club responded to his plea. He said, "I later learned that the Sierra Club had scheduled a helicopter trip to Anacapa." Ultimately Governor Ronald Reagan interceded and prohibited the California Department of Fish and Game from landing helicopters on Anacapa, and Superintendent Don Robinson prohibited anyone from going ashore until the late summer of 1971.

In an article entitled "The Silent Shore," in *Audubon Magazine,* November, 1971 Faith McNulty wrote the following in reference to the California Coast: "This year and for the past several years, brown pelicans and cormorants have failed to reproduce. On the offshore islands where thousands of these birds annually gather in noisy, busy, breeding communities, biologists have found almost no young; only abandoned nests and scattered eggshells. On the Farallon Islands, 30 miles west of San Francisco, where thousands of murres and ashy petrels nest, eggshells are measurably thinner than normal, giving warning that like the pelicans and cormorants, these birds, too, may reach the point where reproduction is impossible. . . . It is reasonable to surmise that any birds that linger and feed in the contaminated waters off California will be affected by the high level of persistent pesticides. . . . Water and air are the two most mobile components of the environment, and chlorinated hydrocarbons such as DDT travel freely along with them. . . . No one can predict if the pelican will ever again nest successfully off of California. It depends on whether there is a sufficient decline in pesticide levels and if that decline occurs while there are still brown pelicans to repopulate the old colonies." Miss McNulty said that Dr. Risebrough predicts it will be twenty to thirty years before the California Coast is healthy for fish-eating birds.

In 1970, television and movie actor Eddie Albert joined Dr. Risebrough on a trip to Anacapa. Albert said, "I guess that's when I got mad, really fighting mad. We climbed the bluffs to where the pelicans had flown since time began. There were thousands of nests in a rookery that ordinarily produced more than 1,000 chicks. What an awful sight it was. The eggs were there but they were all collapsed, 10,000 and more of them." Albert concluded, "What had happened was that fish had consumed the DDT that blows quite gently out from the farms, ranches and fields all over Western America and down into the Pacific." He said that the birds had fed upon these fish and had consequently laid weak-shelled eggs. He added, "What I saw that day is not only happening to the pelicans but also to the cormorant eggs, and seagull eggs; it's happening to the osprey, the petrel and on land to the peregrine falcon and eagles—I could go on but what it all spells out is extinction. The end of the line."

In December, 1971, Dr. Edwards completed a comprehensive report on the 1971 California Brown Pelican census. According to Dr. Edwards, approximately 1,500 brown pelicans roosted or nested on Anacapa with a conservative estimate of 200 young born in the spring. He estimated that 20,000 pelicans inhabited other nearby islands with nearly 1,000 pairs nested on San Nicolas alone. Approximately 7,000 adults and juveniles were observed on the Farallon Islands off San Francisco that year. He said, "There are at this time approximately 30,000 to 40,000 brown pelicans gliding along the California Coast to the surprise of the residents who believed that this bird was nearly extinct and to the consternation of the anti-DDT activist whose propaganda depended upon concealing the truth from the public and the press."

Howard Leach, special wildlife investigator for the California Department of Fish and Game, contended that these birds had actually flown up from Mexico. He said that his department would investigate the fact that "downy" baby pelicans had been observed in the San Nicolas nests. "If they saw downy birds in the nests there, they could be 100 percent right that the pelicans are breeding once again in the Channel Islands. Birds that small couldn't swim there from Mexico, but if they didn't see downy heads they could be 100 percent wrong." It is certainly hoped that when the Department of Fish and Game investigates the Edwards report (the data were witnessed by Dr. Russell Stodd of Portland, Oregon) it will approach the nesting grounds with proper caution and not duplicate the helicopter landings and the intrusions of Gress on Anacapa.

Jim Fowler, a zoologist, was a guest on the Dick Cavett Show, July 8, 1971. With him he had two caged brown pelicans and a huge snapping turtle. First Fowler showed the audience the two panic-stricken pelicans from the Florida area and explained that these birds were rapidly disappearing because of DDT in the ocean. He stated, "They have disappeared in California already." Fowler informed the audience that the pelican is at the top of the food chain. He then recaged the pelicans and displayed the snapping turtle, which incidently is also at the top of the food chain and lives in the same area as the Florida brown pelican. Fowler said, "They [referring to the turtles] are not disappearing. They are unbeneficial as far as man is concerned because they feed off the birds that man shoots."

In the *Brown Pelican News Letter,* November, 1968, the brown pelican population is described as follows: "The population of pelicans in Florida is still apparently in good shape. Lovett Williams and Larry Martin of the Game and Fresh Water Fish Commission made an aerial survey of both coasts on 6 and 7 of May and found 24 active colonies containing a total of 6,705 *pairs* of birds. The pelicans in the Lower Keys and Florida Bay had largely finished nesting at this time and estimates by John Ogden of the National Park Service and Jack Watson of the Bureau of Sport Fisheries and Wildlife indicate that another 1,775 pairs probably should be added for a total population in the neighborhood of 8,500 *pairs.*" [Emphasis added.]

Dr. Robert Finlay, Jr., officially representing the Denver Laboratory of the United States Bureau of Sport Fisheries and Wildlife, testified before the California Water Quality Control Board in Los Angeles on February 20, 1971. He stated, "50,000 brown pelicans disappeared from the Texas and Louisiana coast since 1961." Yet in the Poage Hearings of 1971 it was brought forth that the total number of adult brown pelicans along the entire Texas coast in 1942 did not exceed 300. Dr. Finlay's statement that 50,000 pelicans had disappeared since 1961 has been widely quoted and used as a basis for claiming a decline in the number of brown pelicans. If there were only 300 birds in 1942, and 50,000 birds have *disappeared* since 1961, and there are *still* pelicans in that area, instead of showing a decline, Dr. Finlay has probably reported the most fantastic population explosion of birds over a twenty-year period in history. Moreover, this "population explosion" occurred at the height of DDT's usage.

DDT was used extensively for the first time in and around Naples, Italy. More DDT has been used in that relatively small country than has ever been used in any comparable area in the United States. Researchers should go to Italy for an extensive study of bird life there. Meanwhile, it is dangerous to claim that DDT causes thin eggshells. Birds have indeed been known to lay thin-shelled eggs, even before the advent of DDT, but allowing the environmentalists to place the blame on DDT in order to advance their propaganda campaign may well discourage scientists from pursuing the actual cause or causes, and ultimately result in a *real* decline of bird populations.

Lakes, Rivers, and Oceans

7 Pollution of our oceans, inland streams, and lakes is a serious and complex problem that has been exploited for many reasons, including direct or indirect financial gain. Because it is an acknowledged fact that in a few isolated instances in the 1950s excessive amounts of DDT spray resulted in fish kills, the doomsday lobby has added the contamination of fish by DDT to its list of reasons for demanding that this pesticide be banned. Neither of the lobbyists' two prime targets—detergents and DDT—is the major cause of fish kills. According to the Federal Water Pollution Control Administration, the two primary water contaminants responsible for fish kills are industrial wastes and manure.

Virtually everything eventually finds its way into our oceans as it is carried there by such means as run-off, irrigation water, and rain. The Continental Divide determines the direction of the flow. DDT is no exception; however, its significance to our waterways and fish has been exaggerated beyond all reason. It is but one single material among many thousands. The following table shows that the natural fertilizer manure has been responsible for nearly four times as many fish kills as *all* pesticides *and* chemical fertilizers combined.

POLLUTION CAUSING FISH KILL IN THE U.S. IN 1967.

Source of Pollution	Reported Fish Killed	% of Total
Industrial wastes	8,087,091	72.7
Municipal wastes	643,304	5.8
Transportation wastes	143,123	1.3
Other operations	638,266	5.7
Insecticides, poisons, etc.	329,130	3.0
Fertilizers	10,000	0.1
Manure-silage drainage	1,268,137	11.4
TOTAL	11,119,051	100.0

Source: "Pollution caused fish kills—1967." Federal Water Pollution Control Administration, Washington, D.C. 8:1–16

Despite the fact that fish kills have been reported from a variety of causes, the total effect on the fishing industry and on sport fishing has not altered appreciably the over-all catch in the United States or the world. Certainly there is room for improvement, but there is no reason for panic.

The United States Department of Commerce's publication, *Fisheries of the United States, 1970,* reported, "World fishing, which has more than tripled in volume since 1944, dropped by a little over one per cent in 1969—the first decrease in the postwar period. Until 1969, world catch trends had been uninterruptedly upward ever since the postwar construction of fishing fleets. Although the 1969 catch was slightly below that of 1968 it was double the catch (88.2 billion pounds) taken in 1960."

Despite the persistent claims that DDT is drastically affecting our aquatic life, the Department of Commerce's catch records are hard to ignore. Their records show that in 1940 (prior to DDT usage) the total U.S. catch was 4,060,000 pounds. In 1950 (a year of moderate DDT usage) the total U.S. catch was 4,901,000 pounds. In 1960 at the height of DDT usage, the total U.S. catch was 4,942,000 pounds—the highest over a thirty-year period. In 1970 (minimal DDT usage) the total U.S. catch was 4,884,000 pounds. These figures may be found on page four of *Current Fishery Statistics,* No. 5600, a report that was released in March, 1971, and which may be obtained from the Superintendent of Documents, United States Government Printing Office, Washington, D.C. 20025.

In his testimony before the State of Washington DDT Hearings in October, 1969, William G. Gusey, a wildlife specialist, stated that the extensive publicity given to the Coho salmon's failure to reproduce successfully in Lake Michigan was misleading. This failure was attributed to high DDT residues in Coho eggs. Mr. Gusey pointed out that this widely publicized egg hatch failure had occurred at only *one* of the many hatcheries in Michigan. He said, "All other hatcheries reported normal reproduction of viable fry (newly hatched fish) with fry mortality within acceptable limits for hatchery operations. Certainly the growth and abundance of Coho salmon in Lake Michigan attests to the capability of the fish to thrive in the lake." According to Mr. Gusey, figures from the Great Lakes Fisheries Commission show that in 1967 the production of all species of fish in Lake Michigan was greater than ever recorded from that lake since the first records in 1879. The Commission's report showed that *all* of the Great Lakes fish production in 1967 was 127,365,000 pounds. This figure was exceeded only in 1956 and prior to that not since 1918.

DDT did not create or solve the pollution problem in Lake Michigan, and it is worthwhile to digress briefly to describe the actual problem there and how it was solved.

In his testimony before the Committee on Agriculture, House of Representatives in March, 1971, Dr. Donald A. Spencer, a veteran of thirty-four years' service with the United States Department of the Interior, Fish and Wildlife Service, stated, "The Michigan Department of Natural Resources hatches and rears Coho and chinook salmon in their hatchery system for release into streams tributary to Lake Michigan when they are approximately five inches long and weigh one ounce each; eighteen months later the Coho salmon return to these same streams to spawn having an average weight of ten to twelve pounds. Beat that for production in a lake that is supposed to be so polluted." Dr. Spencer also noted that the lake-running steelhead, brown, and rainbow trout are making spectacular growth and reproducing naturally in Lake Michigan.

The sea lamprey had virtually decimated the fishing industry of the Great Lakes in the 1940s and 1950s. The sea lamprey (lamprey eel) is approximately 18 inches long. This parasitic killer attaches its sucker-like mouth to the bodies of lake trout, whitefish, and salmon and sucks out the blood and other vital body fluids of the host. A lamprey can kill a large trout within a single day. Frequently several lampreys will attach themselves to one fish.

When the lampreys entered the Great Lakes through the Welland Canal, which by-passes Niagara Falls, their amazing reproductive capacity and ability to breed in fresh water made *them* the number one problem of the Great Lakes fish industry.

Sometimes fish themselves pollute a body of water, as did the alewives which, along with the lampreys, came from the sea to invade the Great Lakes. Because the lampreys killed many of the lake trout which fed upon the young alewives, the alewives, too small to attract the lamprey, thrived in Lake Michigan. When the alewives reach seven to eight inches in length, they die—usually in great numbers at a single time. Thus a severe pollution problem resulted in 1967 when millions of these fish died in the lake that year.

George Leposky, in an article entitled "The Alewife Invasion," in *Animal Kingdom,* December, 1969, wrote, "Chicago spent nearly a quarter of a million dollars or about $4,500 a day throughout much of the summer to clear its beaches of dead alewives. In Gary, Indiana, the cost was $75,000 and similar sums were spent by other cities around the lake. Factories were shut down when dead fish, drawn through water intake pipes, clogged machinery. Owners of summer homes near the shore were driven away by the odor of decomposing alewives, and the West Michigan Tourist Association estimated that resort owners lost more than $50 million worth of business."

The state of Michigan solved the alewife problem effectively by stocking the lake with approximately three and one-half million Coho salmon and one and one-half million chinook salmon. Since salmon eat alewives, the salmon have thrived, thus controlling the alewife population while at the same time restoring commercial and sport fishing to Lake Michigan.

Equally effective for the lamprey eel was the pesticide, TFM, which kills the larval sea lamprey without harming commercial and sports fish. The TFM program began in 1958 and applications of TFM are used regularly to prevent re-establishment of the lamprey. This pesticide was first used in the tributaries of Lake Superior and, because of its remarkable success, it is now used for lamprey control in all of the Great Lakes.

Dr. Spencer told the House Agriculture Committee, "The Department of Natural Resources calls Lake Michigan the world's greatest fishing hole, and they have every right to say that now. All this is possible because of a pesticide chemical, TFM, which is employed to keep the lamprey eel under control. If the pressure is relaxed on this parasitic eel the entire fishery will go down the drain."

As a fringe benefit, TFM also kills "rough" fish—noncommercial and nonsport fish—when used in concentrations exceeding 3 ppm and below 7 ppm. These examples of the lamprey and alewife show how the combined efforts of man, a pesticide, and nature solved two very complex problems in Lake Michigan.

The following headline appeared in the *San Francisco Chronicle,* June 14, 1971: "Baby Fish Dying from DDT in the Sea." Under the headline was the following lead paragraph: "Nearly 25 per cent of all the DDT manufactured to date is now in the world's oceans where it is killing baby fish, according to a report by a panel of the National Academy of Sciences." According to Dr. Gunther, "The entire story is a gross distortion based upon inadequate analytical evidence and extrapolation to completely unrealistic extremes." Nevertheless, the statement of the National Academy of Science, which was released by the *Washington Post News Service,* appeared in newspapers throughout the United States.

Professor John D. Furrer, of the University of Nebraska, has calculated the length of time it would take to saturate the oceans of the world with DDT. He based his calculations on three deliberately exaggerated assumptions: (1) that 300 million pounds of DDT are produced in the world each year (actually closer to two-thirds of this amount is produced annually); (2) that no metabolic breakdown would occur (an absolute impossibility); (3) that all of the DDT that has *ever* been produced had been purposely dumped into the oceans.

Dr. Furrer concluded from this study that it would take over 9,000 years to reach a concentration of one part per billion of DDT in all the oceans. Therefore, if 25 percent of the DDT produced to date were actually in the world's oceans, in approximately 9,000 years we would have a concentration of 0.25 ppb of DDT. Without a closer look at the facts and figures, the statement that 25 percent of all the DDT ever made is in the world's oceans does indeed sound ominous.

Within the news article just discussed there was a paragraph that illustrates a vital point in the anti-DDT campaign. The public has been led to believe that DDT creates a serious danger to sea life. The paragraph read: "The ripe eggs of speckled sea trout on the South Texas Coast now contain some 8 parts per million of DDT residue, and the area's speckled sea trout have dropped from 30 per acre in 1964 to .2 per acre in 1969. Five parts of DDT [sic] residue causes 100 percent mortality in young fresh water trout."

The article, written in 1971, covered the period between 1964 and 1969. There was no mention of the three devastating hurricanes which struck the Texas coastal area in the period—hurricanes that churned the water in the spawning beds of this estuary where sea trout spawn and the eggs hatch and develop. The article did *not* say that *even at the time it was written* sea trout numbers had returned to their previous numbers.

An entirely different story came forth when Dr. Philip A. Butler, Director of the United States Gulf Breeze Laboratory in Florida, testified under oath before Hearing Examiner Edmund M. Sweeney at the Environmental Protection Agency's DDT Hearings in Washington, D.C., October 28, 1971. Dr. Butler was questioned regarding the level of DDT in sea trout eggs. Dr. Butler stated that he had served as a member of the committee which prepared a report entitled "Chlorinated Hydrocarbons in the Marine Environment" for the National Academy of Science. Dr. Butler said that he had supplied rough data and background information for the report.

The following statement from the report was then read to Dr. Butler: "In speckled sea trout on the South Texas Coast, DDT residues in the ripe eggs are about 8 parts per million. This level may be compared with the residues of 5 parts per million in fresh water trout that causes 100 percent failure in the development of sack fry of young fish."

Dr. Butler verified that the quoted paragraph was based on data that he had supplied and that this data was generated by the Texas Park and Wildlife Department. When he was asked how many of his results in sea trout eggs exceeded 4 ppm of DDT, his answer was, "One."

Dr. Butler was then asked, based on the material he had just verified, if the reading would be considerably less than 4 ppm and more like one ppm. Dr. Butler ultimately agreed that an average reading would be closer to one ppm.

The point intended in relating this testimony is not to quibble with the small difference between one ppm and 8 ppm. Since the Food and Drug Administration's tolerance level for DDT in fish is 5-ppm, Dr. Butler's implication that sea trout eggs contained levels of 8 ppm *is* significant to the fisherman and further is an indication that Dr. Butler, a member of the Environmental Defense Fund, presented a distorted report, using the single highest reading rather than the average or the mean. Thus, any conclusions based on this data are scientifically meaningless.

In April, 1969, the Food and Drug Administration imposed an interim limit of a 5-ppm tolerance level of DDT for interstate shipment of fish. For years the DDT tolerance level in virtually all foods was 7 ppm and, as has been shown, there were no ill effects. Since fish makes up only a very small part of our overall diet, why would 6 or 7 ppm of DDT in a fish be injurious when 7 ppm of DDT in other foods was not?

The tolerance level of 5 ppm set for fish is an arbitrary and unrealistic figure. The regulation that fish containing more than 5 ppm are banned from interstate shipment is absurd. The impression is that a fish caught in Texas containing 6 ppm of DDT could be safely eaten by a Texan, but if the same fish were shipped across interstate lines to Arizona, an Arizonan would suffer deleterious effects from eating it.

George M. Woodwell, one of the three original founders of the Environmental Defense Fund, wrote an article entitled "Toxic Substances in Ecological Cycles" in the *Scientific American,* March, 1967, in which he stated: "Because of the wide distribution of DDT the effects of the substance on a species of animal can be more damaging than hunting or the elimination of a habitat (through an operation such as dredging marshes). DDT affects the entire species rather than a single population and may well wipe out the species by eliminating reproduction."

Woodwell does not name any species that has been eliminated by DDT, nor, in fact, could he. According to Professor Walter Ebeling, University of California, Los Angeles: "There are no doubt some large predators such as the Grizzly bear and wolf that have suffered from the encroachment of man. However, *in general,* wildlife is increasing everywhere in the United States. Wildlife is favored by man's agricultural practices, which provide seed, grasses, and impoundment of water for waterfowl, particularly where barren desert areas are made to bloom. Fish populations are thriving in most of our aquatic habitats. In the suburbs of Los Angeles, *within the city limits,* such wildlife species as deer, rabbits, raccoons, opossums, foxes, skunks, and birds are abundant. Where has the spring become silent?"

The statement by Dr. Ebeling appeared in the *Los Angeles Times,* November 19, 1971. That newspaper had consistently printed article after article protesting the use of DDT, but on November 28, the *Times* reversed its stand in an editorial which read in part, "This newspaper was itself persuaded for a while that DDT ought to be banned altogether. But we, like the 'hysterical environmentalists' Dr. Borlaug attacked, had not sufficiently considered its importance to the underdeveloped nations nor the possibility that the alternative pesticides might be more dangerous to life." For a newspaper of the size and prestige of the *Los Angeles Times* to make such a statement, and to admit an error, is indeed encouraging and noteworthy.

Dr. Donald Spencer stated at the 1971 House Congressional Hearings, "There have been accidents and misuse of pesticides which have resulted in kills of fish and wildlife. In practice these cases have almost invariably been followed up and corrective measures have been instituted."

In her chapter "Rivers of Death" in *Silent Spring,* Rachel Carson wrote at great length about the fate of the Atlantic salmon in the Miramichi River in New Brunswick, Canada. That DDT was responsible for a heavy fish kill in this relatively small river is acknowledged. However, Miss Carson's version of the story does not square with the facts.

New Brunswick, a maritime province on Canada's Atlantic seaboard, relies heavily on forestry and forest by-products for its economy. A little over 85 percent of the province is forested primarily with spruce and balsam fir.

The spruce budworm, though omnipresent in the area in small numbers, can and does reach sudden catastrophic populations. The destructive capacity of these insects is overwhelming. They are able to defoliate an entire forest in only a few months. If this destruction is not halted before a second defoliation, virtually the entire forest would be destroyed. Widespread infestations of the spruce budworm erupted in the New Brunswick forests in the 1950s. The provincial government ordered aerial spraying of DDT to protect the province's main industry, forestry. Unfortunately, too much DDT was used in 1954, and a considerable amount fell into the Miramichi River. The forests were saved but the salmon suffered heavy losses.

According to Miss Carson, "By August not one of the young salmon that had emerged from the gravel beds that spring remained. A whole year's spawning had come to nothing. The older young, those hatched a year or more earlier, fared only slightly better. For every six young of the 1953 hatch that had foraged in the stream as the planes approached, only one remained. Young salmon of the 1952 hatch, almost ready to go to sea, lost a third of their numbers."

Miss Carson did not tell the full story. Heavy salmon declines had been reported in the Miramichi since 1930; thus the provincial government set up electroseining stations on the northwest Miramichi in 1950. (Electroseining is a means of stunning fish by electric shock, thus enabling scientists to net, examine and count the fish in a given area. By using this safe method, fishery biologists are able to determine the age, sex, size and number of fish in a given location at a given time.)

Miss Carson did not report that the count of fry during the three years preceding the use of DDT was 717 while the total fry count during the three years after DDT was used was 2,755.

The parr (two- to three-year-old salmon) numbered 3,940 during the three years prior to the spraying, compared to 3,287 during the three years after spraying. Information regarding the Miramichi salmon count was provided by the Canadian Fisheries Board, which has kept annual records since 1951.

Miss Carson was aware of the fact that unprecedented numbers of salmon fry were observed in the Miramichi in 1955. In 1953, 359 fry were counted and in 1955, the year *after* DDT's use, an all time high of 1,943 fry were counted. On pages 133–134 she attributes this phenomenon to Hurricane Edna, which, according to Miss Carson, "brought torrential rains to the New England and Canadian Coasts. The resulting freshets carried streams of fresh water far out to sea and drew in unusual numbers of salmon." It is true that some fish get tossed about by hurricanes and become stranded; torrential rains, however, do not *draw in* salmon to spawn. Salmon return to their native stream to spawn *when they are biologically able to do so.* They do not return to their native stream from the ocean for any other reason *except* to spawn.

The young Atlantic salmon traditionally stay in their parent stream three and one-half years: one year as fry (newly hatched salmon), one year as small parr, and one and one-half years as large parr. At the end of this time, they move downstream to the ocean where they remain for from one to two years before returning to their parent stream to spawn. An Atlantic salmon (unlike a Pacific salmon) must be at least four years and usually five years old before it has the *reproductive capability* to spawn. The sudden occurrence of a hurricane would not alter this centuries-old biological reproductive pattern.

The 1962 figures from the Fisheries Research Board of Canada show that there was indeed a sharp reduction in the number of fish the year DDT was sprayed; however, there were enough survivors to form a basis for continued reproduction in later years. Even with DDT spraying, pulpwood blockage in some of the spawning streams in 1958, and pollution by a base metal mine in 1961, the number of fry, small parr and large parr remained surprisingly constant. Thus the salmon were able to survive all of these hazards and still reproduce. The Canadian Fisheries Board's report shows that angling catches on the Miramichi between 1961 and 1967 increased 214 percent while commerical catches increased 225 percent. Miss Carson's title "Rivers of Death" implied a state of permanence rather than a temporary condition.

In a speech at Rutgers University, November 18, 1969, Dr. Robert White-Stevens, Director of the Bureau of Conservation and Environmental Science at Rutgers, said, "There have undoubtedly been fish kills along inland streams and waterways that are, in small part at least, attributable to pesticides which enter the water either by direct application or by erosion of treated nearby soils. In a few cases the irresponsible dumping of residual sprays, old containers, etc., directly into streams and ditches has killed fish." Dr. White-Stevens explained that the published figures of unnatural fish kills in the United States over the past seven years show that only one to three percent of the total fish kill was caused by pesticides. DDT is only one of many pesticides. He concluded, "When the decision rests between producing food or fiber for our people and the protection of millions of acres of forest and billions of board feet of lumber or alternatively preserving less than 0.1 percent of our fresh water fish, which rapidly reproduce themselves anyway, there really is no question as to which choice should be made." Granted that it is difficult to make such decisions, but surely it would be irresponsible on the part of those, in whose hands such decisions must rest, *not* to make them.

The alleged effects of DDT on clams and oysters are discussed on page 151 of *Silent Spring*. Miss Carson wrote, "Oysters and clams may concentrate these poisons [insecticides] in their digestive organs and other tissues. Both types of shellfish are normally eaten whole and sometimes raw. Dr. Philip Butler, Bureau of Commerical Fisheries, has pointed out an ominous parallel in that we may find ourselves in the same situation as the robin. The robins, he reminds us, did not die as a direct result of the spraying of DDT. They died because they had eaten earthworms that had already concentrated the pesticide in their tissues."

As has been shown (Chapter Six), the robin population in the United States increased 1137 percent between 1941 and 1960. In light of this, to "find ourselves in the same situation as the robins" is indeed ominous.

In his testimony before the Environmental Protection Agency Hearings, the same Dr. Philip Butler said that in more than 8,000 samples of oysters tested, 65 percent had been contaminated with DDT or its metabolites. He said that the highest level of DDT was 5.3 ppm. The same information appeared in an article by Dr. Butler, "Monitoring Pesticide Pollution," *Bio-Science*, 1969, and the 5.3 ppm figure has been widely quoted.

Under cross-examination, Dr. Butler stated that the figure in excess of 5 ppm (the FDA tolerance level) was the result of a *single isolated instance* of gross pollution and that actually the national average is less than 0.1 ppm and only rarely as high as 0.5 ppm. Dr. Butler said that the slight decrease in oyster populations had been due to *disease*.

According to the United States Department of Commerce, *Current Fisheries Statistics*, No. 5600, March, 1971, "The 1970 harvest of oyster meats was 50.6 million pounds, compared with 51.9 million pounds in 1969. Production in and around Chesapeake Bay and the state of Washington was a little better than in 1969, but declined in South Carolina and all of the Gulf States. A factor in the decline in the Gulf States was that oyster reefs in Louisiana and Mississippi had not fully recovered from the effects of Hurricane Camille that devastated the Mississippi Coast and the lower reaches of Louisiana in August, 1969."

In his testimony before the U.S. House of Representatives'

Pesticide Hearing, March, 1971, Dr. Donald A. Spencer, wildlife specialist, said that the waters around Long Island Sound
are as polluted as any in the United States. He explained that
through careful management, proper controls, and the use of
oyster hatcheries, 250,000 bushels of bluepoint oysters were
harvested in Long Island Sound in 1970, and that he expected
a harvest of 500,000 bushels in 1972. Dr. Spencer said that in
spite of the fact that Chesapeake Bay has pollution problems
(approximately eleven million people live in that drainage ba-
sin), today over 600 million pounds of fish and shellfish for our
tables are harvested there. He pointed out that this catch is
higher than at any time since 1880 when records were first kept.

It is certainly true that some rivers and lakes are heavily
contaminated from many sources and with many substances,
but progress *is* being made. The early problems of DDT fish
kills have been eliminated by careful avoidance of the spraying
of DDT directly on streams and by careful control of the
amount used.

Despite the prophets of doom, our rivers are, in fact, rivers
of life, and our oceans remain bountiful, and man, refusing to
accept defeat, is gradually winning the battle of water pollution.
Since pollution did not "happen" in a day, we cannot expect
this complex problem to be solved "overnight."

The
Web of
Life

8 The accumulation of DDT in the fatty tissues of human beings has been exaggerated to such a degree that many people actually fear for their personal safety. The average world citizen ingests through his diet no less than one and no more than five grams of DDT in his entire lifetime. In the United States the average is approximately one gram (1/454 of a pound) for a seventy-year life span. Five grams of DDT is fractionally less than a level teaspoon. Therefore, the average American ingests about one-fifth of a teaspoon of DDT *in his entire life.* These figures are based on the average amount of DDT ingested between 1950 and 1970 and on the projection that this level would continue for a seventy-year life span. The level is, in fact, decreasing, because the use of the pesticide is decreasing.

DDT, like vitamins A, D, and K, is stored in the fatty tissues of the body until it is metabolized. The DDT in the body is slowly but constantly being broken down by the liver and is replaced in decreasing amounts in our daily diet. At any given time, the average American has approximately five parts per million in his body fat, an amount in his entire body which is too small to be detected by the human eye. By contrast, lead and strontium 90 (toxic elements unavoidably ingested and stored in the human body) are not metabolized but are retained in the bones for life.

Frequently we hear that DDT is laid down in human fat forever. This is impossible since human fat has a 100 percent turnover rate in less than a year. The half-life of a pesticide is the amount of time required for half of the original amount to be broken down or otherwise disappear. According to clinical toxicologist Dr. Richard T. Rappolt, Sr., "The half-life of DDT in fat tissue is approximately three to six months."

Professor John D. Furrer of the University of Nebraska has stated, "Charges that the public is being poisoned by DDT in its food and drink are without foundation." Dr. Furrer has also pointed out that the total DDT intake of a 154-pound man during a seventy-year life span would be only 1.28 grams. To recall Dr. Rappolt's treatment of barbiturate overdoses in San Francisco (Chapter Five), 5 grams of DDT were given in a single dose to each of three women in the doctor's successful attempt to save their lives.

Dr. Wayland J. Hayes, Jr., of the Vanderbilt School of Medicine, in his article "Monitoring of Food and People for Pesticide Content" in *Scientific Aspects of Pest Control,* 1966, wrote, "We have found that the loss of DDT from storage, although always slow (less than 0.3 percent per day in man), is always more rapid when the storage is high than when it is low. Thus, the rate of excretion of DDT changes with the concentration in the fat. The important principle is that all compounds reach or at least approach a steady state of storage provided the dosage is constant and is tolerated by the organism."

Dr. Hayes also stated that since thousands of DDT factory workers and hundreds of thousands of spraymen have now been exposed to extremely high levels of DDT for more than one-eighth of their life-span and have not suffered detectable harm, it is most unlikely that the general population would be harmed by dosages 200 to 970 times *smaller.*

It has been shown by Dr. Hayes' study that although traces

of DDT are found in virtually everyone in the general population of the United States and other countries here has been *no* increase in the storage of DDT by the general population since it was first measured in the United States in 1950. DDT is not cumulative. If the amount of DDT taken in decreases, the DDT in the fat tissues decreases, and if the DDT taken in increases, the DDT in the fat tissues increases.

It must be remembered, however, that even when extremely high amounts of DDT were present in human fat tissues, no harm to the health of those examined was reported. According to Dr. Francis Gunther, "Dr. Hayes has shown that the storage level of DDT does not increase indefinitely during continued exposure. It reaches a plateau; however, that level will decline if the level of DDT taken in is decreased. All this fuss about DDT building up indefinitely in the fat tissue is utter nonsense."

In 1969, in an experiment by scientists of the United States Department of the Interior at the Gulf Breeze Laboratory in Florida, the decomposition rates of four pesticides and their metabolites were tested in sea water. This was done by placing, into clear, one-gallon, glass jars of sea water, a concentration of each pesticide equal to three parts per billion. The jars were then sealed and immersed in an outdoor flowing sea water tank for forty days. The results showed that DDT and its principal metabolites, DDE and DDD, decompose rapidly in sea water. It is interesting to note the decomposition rate of parathion, the insecticide that Dr. Charles F. Wurster, Jr., and many others claimed to be *short-lived*. It is *more* persistent in sea water than the combined total of DDT and all of its metabolites. The accompanying 1970 table from the Department of the Interior shows the results of this experiment.

After 38 days the decomposition was 94.5 percent for DDT, 95.1 percent for DDE, while DDD (TDE) which appeared in very minute traces on the 17th day decomposed by 55 percent.

Stability of Pesticides in Natural Sea Water
(Salinity 29.8 ppt pH 8.1)

Days after start of experiment

Pesticide	0	6	17	24	31	38
p.p′-DDT	2.9	0.75	1.0	0.27	0.18	0.16
p.p′-DDE*	0.0	0.096	0.95	0.065	0.034	0.037
p.p′-DDD*	0.0	—	0.081	0.041	0.038	0.037
Aldrin**	2.6	0.58	0.096	<0.01	<0.01	<0.01
Dieldrin*	0.0	0.74	1.0	1.0	0.75	0.56
Malathion	3.0	<0.2	<0.2	—	—	—
Parathion	2.9	1.9	1.25	1.0	0.71	0.37

(All readings are in parts per billion)
*Metabolites of parent compound.
**From the seventeenth day onward, two unidentified peaks appeared on the gas chromatographic charts after aldrin had eluted.
<Indicates "less than."

Although 99.6 percent of the aldrin had decomposed by the end of the experiment, only 44 percent of the dieldrin, the metabolite of aldrin, had broken down. The breakdown of parathion was 87.3 percent.

The experiment shows that DDT and its metabolites rapidly break down in sea water, contrary to the ridiculous exaggeration that 25 percent of all the DDT ever manufactured is still in the oceans (see page 95).

In her article, "The Silent Shore," *Audubon Magazine,* November, 1971, Faith McNulty wrote: "In October, 1970, The Environmental Defense Fund decided that it had sufficient evidence that Montrose was polluting the coastal water to ask the court for an injunction. In November two graduate students, working with Dr. Risebrough, took water samples from the sewer below Montrose and found that at least 300 pounds of DDT was flowing through the sewer line daily. At the same time EDF discovered that The Los Angeles County Sanitation District, a municipal body responsible for sewers, had measured the effluent months previously, had found that 600 pounds a day was going through the sewer, and privately asked Montrose to desist."

After explaining that the Montrose Chemical Corporation began its operation in 1947, Miss NcNulty added, "Thereafter, until 1970, the waste products of the plant, laden with DDT, were drained into the sewer and spewed into the Pacific Ocean a few miles away."

According to Miss McNulty, "A reasonable estimate of such losses is one-half of one percent. If Montrose lost that much DDT it would amount to 1,000 pounds of DDT per day washing into the sewer."

The capacity of the Montrose plant is seven and a half million pounds per month. Their average production has been six million pounds per month or approximately two hundred thousand pounds per day, which is equivalent to 34,000 dollars at seventeen cents a pound. One-half of one percent (0.005) of this amount equals 170 dollars per day or 5,100 dollars per month or 61,000 dollars per year alleged effluent loss. Considering the normal overhead of any industrial plant, it is unlikely that any company could withstand the loss of 61,000 dollars per year, literally down the drain.

On December 8, 1971, I received a personal letter from John D. Parkhurst, Chief Engineer and General Manager, County Sanitation District of Los Angeles County. Mr. Parkhurst enclosed a chart entitled "Montrose Chemical Effluent Samples (24 Hour Composites)." Unfortunately the chart contained only three readings of total DDT pounds per day during October and November, 1970, the particular months referred to by Miss McNulty. These readings do, however, serve as noteworthy examples. On Wednesday, October 7, 1970, the total effluent loss *from* Montrose was 0.14 pound of DDT. Thursday, November 12, 1970, showed a loss of 2.08 pounds of DDT. On Tuesday, November 24, 1970, the total loss was 0.37 pound. Each reading was for a 24-hour period on the day cited.

Mr. Parkhurst also enclosed a detailed map of the Montrose area showing the exact location of the sewer lines in the immediate vicinity of the plant and the exact locations of the county's testing stations.

The testing showed that on October 7 the 0.14 pound of DDT referred to above entered the city sewer *from the Montrose plant;* yet at the district's first testing station just *upstream* from Montrose 25.5 pounds of DDT was detected on that same day. How could Montrose have dumped DDT into a sewer line before the line reached the company?

At the first testing station just *below* Montrose, 22.7 pounds of DDT was detected. On October 7, the Sanitation District's reading showed more DDT entering the sewer line *before* the line reached Montrose than they tested in the same line *below* Montrose. Yet Montrose *added* 0.14 pound to the sewer line. This poses a number of interesting questions. Were there unknown factors interfering with the DDT readings in the sewer line? Since PCBs can be mistaken for DDT by analysts who do not take very special precautions, could the PCB run-off from the asphalt pavements and other sources have distorted the sewer readings?

Even if all three readings were added together, the total would be a far cry from the 300, 600, and 1,000 pounds referred to by Miss McNulty.

Widespread publicity has been given to the belief that the continued use of DDT will destroy our ocean's food supply because DDT affects photosynthesis in the phytoplankton of the sea, thus depriving man and animals of needed oxygen. The publicity has been based on Dr. Wurster's article, "DDT Reduces Photosynthesis by Marine Phytoplankton," in *Science,* March, 1968. Dr. Wurster used ethanol (alcohol) instead of sea water as a solvent for DDT in order to achieve his results, which he has stated publicly, but the fact is *rarely* brought to the attention of the public by others. Dr. E. T. York, Jr., Provost of the Institutes of Foods and Agricultural Sciences, University of Florida, analyzed the Wurster experiment in a speech at Camp McQuarrie, Florida, on August 19, 1970:

I am apalled at the manner in which supposedly reputable scientists have accepted the flimsiest sort of evidence as a basis for joining the anti-DDT bandwagon.

Let me illustrate what I mean. A few weeks ago a prominent biologist from one of our prestigious northeastern universities delivered a paper in which he made the following statement: "The DDT in the ocean, even before it enters the food chain, is present in quantities sufficient to impair the photosynthetic efficiency of marine phytoplankton on which the entire aquatic food pyramid is dependent." Of course you readily recognize that if anything were to foul up the photosynthesis of plant life in our oceans, these great bodies of water would ultimately become biologically dead. Furthermore, many have pointed out that the most serious aspect of this would be that man would eventually die from a lack of oxygen which is currently being released into the atmosphere through the photosynthetic activity of marine life. Many scientists and writers in recent months have cited this as perhaps the greatest problem which DDT might cause.

Now let's examine the scientific basis for this concern. It originates from one piece of work reported in *Science* in March, 1968. It involved a laboratory study with five species of marine algae. To the water in which each species was grown, was added different concentrations of DDT in ethanol. Ethanol was used as a solvent because DDT is relatively insoluble in water—in fact, a fully saturated aqueous solution of DDT will have a concentration of only 1.2 parts per billion. The study indicated that algae photosynthesis was substantially reduced with the higher concentrations of DDT (up to 500 parts per billion). However, with a concentration of 1.2 parts per billion—the highest concentration which could be realized in sea water because of the limited solubility of the compound, there was absolutely no effect of DDT on any of the marine algae species. Furthermore, monitoring programs tell us that the actual level of DDT and its metabolites in sea water is one part per trillion or less. Studies also indicate that the concentration of DDT in sea water is not likely to increase above this level even if we continue its usage in public health and agriculture. Therefore, present levels would have to increase one-thousand times or more merely to reach the experimental level where *no effect* on photosynthesis was observed. Furthermore, I would emphasize that there is no conceivable way in which the concentration of DDT in sea water could reach the level in the experiment where photosynthesis was affected in the study.

Now let's go back and take a look at the statement of the biologist which was made as accepted 'gospel' truth: "DDT in the ocean . . . *is present in quantities sufficient* to impair the photosynthetic efficiency of the marine phytoplankton on which the entire food pyramid is dependent." I hope that this illustrates what I meant when I said most of the charges against DDT have been made on the weakest sort of evidence—indeed, much of it will not stand up under the light of dispassionate, objective scientific scrutiny.

In an article entitled "Toxic Substances and Ecological Cycles" in *The Web of Life*, a compendium of biological articles by Love and Love, Dr. George Woodwell, one of the founders of EDF, wrote, "I found in tests in Maine and New Brunswick, where DDT had been sprayed from airplanes to control the spruce budworm in forests, that even in the open, away from trees, about 50 percent of the DDT does not fall to the ground. Instead it is probably dispersed as small crystals in the air. This is true even on days when the air is still and when the low flying planes release the spray only 50 to 100 feet above tree-top levels."

Dr. Woodwell's next paragraph stated, "DDT is an extremely stable compound that breaks down very slowly in the environment. Hence with repeated spraying the residues in the soil or water basins accumulate. Working with Frederick T. Martin of the University of Maine, I found that in a New Brunswick forest where spraying had been discontinued in 1958 the DDT content of the soil increased from half a pound per acre to 1.8 pounds per acre in the three years between 1958 and 1961. Apparently the DDT residues were carried to the ground very slowly on the foliage and decayed very little. The conclusion is that DDT has a long half-life in the trees and soil of a forest, actually in the range of tens of years."

DDT increased from 0.5 pound per acre to 1.8 pounds per acre—a 260 percent increase?

On January 12, 1972, Dr. Woodwell testified at the USDA-Environmental Protection Agency Hearings, Washington, D.C. (These hearings are commonly referred to as the Ruckelshaus Hearings because William D. Ruckelshaus is the Director of the EPA.)

At this time, Dr. Woodwell was questioned about his DDT soil studies in New Brunswick. He stated under cross-examination that his samplings were taken near an airstrip in New Brunswick and that the large amounts of residue which he found there would not be typical in a forest. Dr. Woodwell said, "That's why it had such high levels—high levels of DDT. That's why we picked that site in the New Brunswick."

This helps to explain why he found more DDT in 1961 than in 1958. The DDT did *not* "build up," as Dr. Woodwell reported. The site that he chose for sampling was the spot where pilots tested their spray equipment to be sure that it was functioning properly, and also to calibrate the spray apparatus.

In *The Web of Life,* Dr. Woodwell said: "With my colleagues, Charles F. Wurster, Jr., and Peter A. Isaacson of the State University at Stony Brook, I recently sampled a marsh along the South Shore of Long Island that had been sprayed with DDT for 20 years to control mosquitoes. We found that DDT residues in the upper layer of mud in this marsh ranged up to 32 pounds per acre."

Dr. Woodwell, in his testimony at the Ruckelshaus Hearings, said that the figures for DDT residues in the soil of an extensive salt marsh on the South Shore of Long Island averaged thirteen pounds per acre, with a maximum of thirty-two pounds of DDT per acre. The thirteen pounds of DDT per acre figure was not mentioned in the Love and Love compendium but it has been the most frequently cited figure in other papers regarding the DDT residues in that area. During cross-examination Dr. Woodwell revealed that he and his colleagues took only *six* samples in the salt marsh and *one* in the bay bottom, a grand total of *seven* samples in a 200- to 300-acre salt marsh.

During cross-examination Dr. Woodwell was confronted with the testimony of his colleague Dr. Wurster, who had testified before the State of Washington DDT Hearings in 1969. At that time Dr. Wurster had stated, "We have since sampled that marsh much more extensively and we found that the average, the overall figure on the marsh is closer to *one pound per acre.* The discrepancy was caused by the fact that our initial sampling was in a convenient place and this turned out to be a convenient place for the mosquito commission's spray truck too."

Dr. Woodwell said that he was unaware that his colleague had made the statement but admitted that it was, in fact, true.

When asked if he had ever published a retraction or written a further article giving the results of the later and more extensive samplings, Dr. Woodwell said, "I never felt that this was necessary."

In "The Silent Shore," Faith McNulty stated, "DDE is very slowly degraded. As one scientist remarked, 'We have found no bacteria that would eat it on a platter.' The principal way that it can be lost is if it sinks so deep into the ooze at the bottom of the ocean that it is below the reach of even bottom-dwelling organisms. Once let loose, most chlorinated hydrocarbons remain in the environment for a long time, just how long is still a matter of debate. The reason that chlorinated hydrocarbons are spread around in the first place is that they are poisons."

Miss McNulty also wrote, "The concentration of DDT in organisms depends upon the rate of intake and can rise and fall by amazing multiples. A fish that eats some contaminated food and some uncontaminated food, may only double the concentration in the food supply. But a sea bird that eats nothing but contaminated fish may carry ten times the amount or more. Creatures at the top of the food chain may carry residues at a concentration of more than a million times greater than their environment." More than a *million* times?

Since Miss McNulty has stated that the degree of persistence of DDT in the environment is still unknown, and she does not specify (nor could she) how much DDT is in the environment, the only conclusion that can be drawn from her statements is that creatures at the top of the food chain carry concentrations of DDT more than a million times that of an *unknown* figure for an *unknown* period of time. Her figures and analogies appear spectacular, but they are totally without meaning.

DDT is found at very low concentrations in the algae and small animal life in fresh water and salt water. As organisms are consumed by other organisms in the food chain, the DDT often accumulates or "magnifies" to greater concentrations.

Research has been done in the area of biological magnification but there is no scientific evidence to indicate that a normal accumulation of DDT in fish, animals, or birds causes death or reproductive failure under normal conditions. Accidents have happened. Exaggerated tests have been made. In connection with the Heath experiment that was discussed on pages 75–76, it has to be remembered that the birds in the "control" group were ingesting *normal* amounts of DDT. It would have been impossible to eliminate the minute traces of DDT from their regular diet. Moreover, DDT does not accumulate indefinitely in any living systems. It is constantly being broken down.

Birds can ingest more DDT on a body weight basis than do human beings, simply because birds eat more on a body weight basis than do human beings. Ornithologists estimate that most birds consume amounts of food equal to half their body weight each day. Most of the food is used for energy, and the rest is stored or excreted. Since DDT is more slowly degraded than, for example, carbohydrates and proteins, the concentration of DDT in the body of a bird would exceed that of most other animals including human beings. This is true of all birds—not just eagles, ospreys, pelicans and peregrine falcons.

According to Dr. Francis A. Gunther, "Sunlight (ultraviolet light) and oxygen help to degrade DDT. DDE, an important metabolite of DDT, is also destroyed by sunlight and air. DDT does not decompose significantly as soon as it comes in contact with the atmosphere. Under some field conditions it can remain largely undecomposed for a long period of time. This very persistence enables it to remain biologically effective for a longer period of time, hence avoiding the necessity of repeated application."

In 1956, Dr. Gunther and his colleague Roger Blinn studied the breakdown and half-life of DDT in fruits and vegetables. They found that DDT breaks down quite rapidly on some plants, slowly on other plants, but that it always breaks down. The scientists also learned that when DDT was sprayed on alfalfa in the usual concentrations the initial residue was 13 ppm. After fourteen days, the DDT residue was 5 ppm. Mathematical computation based on these figures shows that the half-life of DDT on alfalfa is eleven days. Thus, at the end of the first eleven days, one-half of the 13 parts per million, or 6.5 ppm, dissipated, leaving 6.5 ppm on the plant. At the end of the next eleven days, half of the remaining 6.5 ppm, or 3.25 ppm, disappeared. At the end of ten half-lives, 99.90 percent of the original DDT was degraded, leaving a residue of 0.013 ppm— an absolutely insignificant amount but detectable by modern techniques. Since some scientists attempt to calculate residues in parts per *trillion,* it is easy to see how minuscule amounts of DDT can be detected after a long period of time even though they are of no significance to either the plant or the consumer.

There is no such thing as an average half-life for DDT, because there are too many variables, such as temperature, moisture, air currents, and a variety of characteristics of leaf and fruit surfaces, of water, and of soil. As has been shown, DDT on alfalfa has a half-life of eleven days. On peach foliage, DDT has a half-life of fourteen to eighteen days, while on the rind of several varieties of citrus fruits the half-life is thirty-three to fifty days. To average these readings would obviously be meaningless.

The anti-pesticide lobby persists in claiming that the use of chemical pesticides, particularly DDT, ultimately places certain species of wildlife in danger of extinction, and by so doing threatens the entire food chain, thereby placing all life in danger. If the elimination of a given species did, in fact, endanger the food chain, life on our planet would have been doomed long before man made his appearance.

The food chain, more accurately called the web of life, is a simple process in which one organism consumes another. Plants, the original source of food, are at the bottom of the food chain. They are eaten by plant eaters (herbivores), which in turn are eaten by meat eaters (carnivores). This process continues through many organisms until the top of the food chain is reached. A creature at the top of the food chain is one that is not normally eaten by other animals.

Many scientists have pointed out that over 90 percent of the plants and animals originally present on this planet are no longer in existence. The same statement would have been accurate two thousand years ago. What these scientists do not always bother to say is that the various forms of plants and animals which became extinct were not all present at *one period of time*. They should always also point out that many of the species which have disappeared have been replaced through the millennia by other species more readily adaptable to their surroundings.

Man has never eliminated a single species of animal through the use of chemicals, with the exception of gunpowder. Despite his efforts, man has never eliminated any insect species, although he has tried very hard. The Carolina parakeet was forced into extinction because man shot it for its plumage. The passenger pigeon, the auk, and the dodo became extinct because they were killed for food.

Many different species of large animals existed on the North American continent fifteen thousand years ago, including elephants, camels, several species of buffalo, sloths, and the horse. Strangely, the horse, which could run faster, ate less, and had better vision than the buffalo, became extinct here and was not reintroduced until the Spaniards conquered Mexico. The camels, elephants, and sloths never returned; yet obviously life continued on the North American continent.

Throughout the history of man, change has been the order of the day. Nature is changing continuously as old species are eliminated and new ones appear. There is no such thing as a status quo in nature. There is no balance of nature. It is a momentary thing.

Walter Ebeling, Professor of Entomology at the University of California, Los Angeles, pointed out the fallacies of nature in balance in an article, "The Public Speaks Out—Nation's No. 1 Pollutant is the Anti-Science Cult," in the *Los Angeles Times,* November 19, 1971. Professor Ebeling wrote, "For those wishing to live in a 'balance of nature' there are still large areas available such as I have seen in Eastern Peru and the Congo. There they might pass their brief lives, developing compost piles and 'organic' gardens to their heart's content, incessantly surrounded by swarms of insects, many of them vectors of disease, some venomous, all pestiferous. These swarms of insects are in balance with nature, including man, and under 'biological control.' "

Dr. Ebeling went on to say, "The dominance of man, and his ability to survive in his present numbers has been the result of his success in bringing about an *imbalance* of nature—in his favor." He emphasized that, unless we wish to decrease our present population drastically and accept modes of life that are still found in some parts of the world, this imbalance must be continued.

I certainly *do not* advocate the elimination of any species and *strongly* urge that, whenever possible, any endangered species of wildlife be protected. Yet man must continue to strive for that favorable imbalance of nature if he is to exist. Such an imbalance can be achieved and maintained only through the judicious use of pesticides and fertilizers and the multitude of other chemicals we strew throughout the world.

Endangered Crops

9 Crops and forests in the United States are attacked by approximately 600 species of insects. Unless most of these insects are controlled by chemical pesticides the production of food and fibers in the United States will be insufficient to sustain the population of this nation and there will be no hope of exporting them to needy countries. The effect of such a situation on those countries who rely on our exported crops would be disastrous and the economic ramifications in the United States would be grave.

The problem of insect destruction of food and fibers is nothing new. In the *Old Testament,* the Prophet Joel lamented the problems of the ancient Hebrews: "That which the palmerworm hath left hath the locust eaten; and that which the locust hath left hath the cankerworm eaten; and that which the cankerworm hath left hath the caterpillar eaten."

Dr. Virgil N. Freed, chairman of the Department of Agricultural Chemistry at Oregon State University, testified at the Washington State DDT Hearings (commonly referred to as the Seattle Hearings) in October, 1969: "Since the beginning of time man was at the mercies of the elements of the environment and the pests that affected his crops and his person. The rise of science in the last century has begun to provide a means by which man could alter his environment, at least to a limited extent, and control the pests that have beset him and his crops. Within the past quarter of a century, a number of chemicals have become available to man, not only for pest control, but for a wide variety of other uses. These were seized upon by society and put into wide-scale practice."

Dr. Freed pointed out that it is often essential to achieve pest control *rapidly* to protect a major crop from *quick* destruction by a *sudden* infestation of insects. Pesticides are the *only* method of preventing widespread destruction because other control methods *presently available* are too slow and too ineffective. In addition, pesticide chemicals allow for production of high yields of quality crops, maximum production at optimal cost, and storage with very little loss and with low filth content.

There are some insect problems in the Northwest for which there is no satisfactory method of control other than DDT or related chlorinated hydrocarbons. Among the problems Dr. Freed listed were certain soil inhabiting insects, certain pests of turf and lawn, and some insects that attack ornamental and tree crops.

A principal use of DDT is for the control of the cotton bollworm, only a minor pest when there is no significant population of the boll weevil; but when organophosphates are used to control the boll weevil, the populations of those beneficial insects that would normally control the bollworm are reduced to such a degree that the bollworm becomes a major problem.

There is no pesticide as safe and effective in controlling the complex of cotton insect pests as DDT. The major problem lies in North and South Carolina, Georgia, Tennessee, Alabama, Mississippi, Louisiana, Florida, Arkansas, Texas, and California. Annual cotton crop production in the United States, incidentally, is valued at 1.2 billion dollars.

When the use of DDT was discontinued on the cotton crop

in Texas, the Tampico area alone reported that cotton acreage which was planted dropped from 500,000 acres in 1967 to 1,200 in 1970. Because the cotton insects had developed resistance to the organophosphates, the farmers found that it was useless to plant cotton.

When Louisiana abandoned DDT for the organophosphates, her cotton loss was approximately 135 pounds per acre, while the drop in California was 250 pounds per acre. The southeastern states face a similar situation with the tobacco budworm.

Oregon, Florida, and New Jersey have suffered a marked reduction in their sweet corn crops as a result of the lack of DDT, while the peanut crops of Georgia and Florida have been destroyed in some cases and the yield seriously reduced in others.

Other crops for which there is no safe and effective substitute for DDT are cabbage, cauliflower, Brussels sprouts, lettuce, tomatoes, potatoes, sweet potatoes, dry, lima and snapbeans, peppers, onions, garlic, alfalfa, and some citrus crops.

In a 1966 speech before Congress, the late Senator Everett Dirksen of Illinois (Senate Minority Leader) discussed the need for agricultural pesticides:

A modern agriculture without pesticides, including the persistent pesticides, is impossible. Any action to ban the use of chlorinated hydrocarbons on the basis of what amounts to superstition, would be a catastrophe to agriculture, a disaster to the consumer, and in view of the world situation, a calamity to the human race . . .

The Fish and Wildlife Service of the Interior and a segment of the Public Health Service persist in issuing some scare reports about pesticides which are misleading. These reports can only confuse, bewilder, and if they are taken seriously, harm everyone concerned . . .

The persistent attacks on pesticides essential for agricultural production and for health, would be ludicrous if they were not harmful and dangerous. There is a steady stream of releases and statements issued from government sources aimed at pesticides. This is a continuing and irresponsible campaign which uses the scare tactic techniques against pesticides and which are becoming increasingly suspect as time goes on . . .

I agree that adequate safeguards must surround the use of pesticides but I am in disagreement with the excessive zeal of publicity-seeking officials whose prejudices are guiding their attitudes and actions . . .

I have never become concerned that, properly used, pesticides propose a health problem. It is time that this whole matter should be looked into from an objective standpoint . . .

Most of the hearings and inquiries held thus far have been directed *against* pesticides. The latter have been on trial. The pesticide industry which makes a tremendous and valuable contribution to our economy has been harassed. I am told that many companies, in view of this situation, have cut back on pesticide research, which is badly needed. The damage done annually by destructive insects must be paid by the consumers. More and more insects become resistant to the pesticides now in use and new chemicals are needed to deal with them. By insisting that pesticides present a tremendous threat to wildlife and human life, Rachel Carson did this country a disservice and her followers are doing still more. I am sorry to say that some officials of the government itself including our Secretary of the Interior, Stewart Udall, are ignoring the evidence and repeating Miss Carson's discredited claims and conjectures. *Silent Spring* is a propaganda tract . . .

Although overzealous in her attack and frequently inaccurate, Rachel Carson did do the world a service by bringing into sharper focus the urgent need for more research to determine the effects of various pesticides on man, plants, and animals.

According to a report in the June, 1967, issue of *The FDA Papers,* an official magazine of the Food and Drug Administration: "American consumers of all ages may be confident that their foods are free from harmful residues due to FDA's tolerance-setting and enforcement procedures. . . . We can conclude that currently the levels of pesticide residues in the Nation's food supply are not approaching dangerous or even alarming levels."

These conclusions were based on the FDA's "Market Basket" studies of the total diet for a seventeen- to nineteen-year-old male consuming a diet level of eighty-two foods over a two-week period. The height of DDT usage in the United States was between 1955 and 1965. This FDA report was released in 1967. The amount of DDT used in this country has steadily decreased since 1965. Further, the FDA tolerance levels for DDT in foods have been decreased to accommodate decreased usage and decreasing residues at harvest. Since the FDA publicly stated that the residues in our foods were safe in 1967, how could they possibly be dangerous at the lower levels of today?

In a speech at California State Polytechnic Institute on February 11, 1972, Dr. Norman E. Borlaug said: "In one-third of the world there is a great affluency today where food is never short, and the people in these privileged positions have forgotten the meaning of 'Give us this day, our daily bread.' They think that bread comes automatically from the supermarket, and it isn't quite so, especially when you look and reflect on the shortage of food in the developing nations where more than half the people live and eke out a livelihood and are all too often short of food."

When Dr. Borlaug introduced a new high yield wheat into Pakistan, wheat production there rose from five to ten bushels per acre to an average of 75 bushels per acre. In his California Polytechnic speech he said, "Without chemical fertilizers and pesticides such as DDT, all this would have been impossible."

Dr. Borlaug also said that the wheat planted in poor lands that have been over-used can produce only six or seven bushels per acre and sometimes as low as three, even with abundant water. "The weeds are so undernourished they can't grow. The plants are so scrawny the rust won't grow on them and the poor insect—he's all emaciated also. When you change the soil you must be in a position to change everything else. Many needed qualities must be developed in the seed such as shortness, high yield and disease resistance.

"The old lands of the world have been depleted of plant nutrients by repeated cultivation, frequently in monocultures over hundreds or even thousands of years." Dr. Borlaug explained that we must find out what nutrients are missing and then supply them for proper plant growth. Usually nitrates and phosphates, and sometimes potash, are present in low amounts and must be added to the soil. As Dr. Borlaug put it, "Under the old system, in depleted soils and worn-out farmlands, hungry and miserable people are in balance with nature—but what a balance."

In the United States, four percent of the population produces enough food to feed the entire nation with surpluses of a variety of foods to export. The United States exported 7.8 billion dollars in food products alone in 1970. Our domestic and export production was achieved on 281 million acres in that year. With the technology available to us in 1940, it would have taken 587 million acres to produce the same amount. Because of our increased population, we do not have that arable 587 million acres.

Until 1973 the average American spent about 16 percent of his take-home pay for food. In Western Europe 35 percent of the take-home pay is used for food, while in Eastern Europe the figure is between 55 and 60 percent. "In India, Pakistan and the Near East, 80 to 90 percent of the family income is spent for food unless there is a famine or a drought in which case all of the money goes for food. In underdeveloped and overpopulated countries, 70 to 80 percent of the people live off of the soil and eke out a bare existence by primitive agriculture."

In summary, Dr. Borlaug stated, "The environment is just one aspect of the problems of man today. We must work on all angles, starting with food. Poverty and unemployment are all tied in with crowding, housing, education, and transportation. The using up of resources, particularly the nonrenewable ones, are all part of this many headed monster we must deal with, and don't ever believe that chemicals are not needed to solve this problem. They are. However, science and technology, without large doses of common sense, are useless."

J. H. Lilly, an entomologist at Iowa State College, in an article, "Soil Insects and Their Control," in the *Annual Review of Entomology,* 1956, said that soil insects do at least some damage to almost every cultivated field in the world, often making the difference between a satisfactory crop and a crop failure. A soil insect lives on or beneath the soil during its growing and feeding stages. Many soil insects are not harmful but some, including white grubs, cutworms, wireworms, rootworms, weevils, and root and seed maggots, are among our worst agricultural pests.

"Anyone who has had much contact with corn culture," Dr. Lilly noted, "can sympathize with the clever pioneer farmer who put into verse the reason why he planted four kernels of corn for every plant he wanted:

> One for the maggot,
> One for the crow,
> One for the cutworm,
> And one to grow."

Drs. Smith and Brierley, Agricultural Research Service, United States Department of Agriculture, Beltsville, Maryland, in a 1956 article "Insect Transmission of Plant Viruses," in the *Annual Review of Entomology,* pointed out that many of our most devastating plant diseases are caused by viruses transmitted by insects. The authors said that if the insect vectors were absent, these diseases would be of little importance; hence, their control is dependent upon the control of the vector insects. An example of a virus disease spread by insects is the Dutch Elm Disease, which has killed many elms in the eastern United States.

The emphasis upon "organically grown food" continues in the United States. Vocal and affluent groups continue to seek a ban on pesticide and fertilizer usage, and urge that we abandon all synthetic aids to society and seek a balance of nature. The same groups zealously strive to promote organic food. *All* food is organic as compared to inorganic, a term which refers to noncarbon compounds such as minerals, salts, and clay. Health food stores misuse the term "organic food." They are actually referring to "organically grown" food, which has come to mean that no synthetic chemicals were applied to the crop or soil. As has been shown, the natural fertilizer (manure) caused 11.4 percent of the fish kills in this country in a sample year (1967) while synthetic fertilizers were responsible for only 0.1 percent of the fish kills (see page 91).

It is a simple process for the home gardener to hand remove the tomato worms from a few tomato plants in his backyard, but millions of Americans live in crowded cities and apartments where home gardening is impossible. They must rely on field-grown tomatoes. The farmer who grows many acres of tomatoes for canneries and food markets could not conceivably use the hand removal technique. He must rely on pesticides, or the tomato industry would be phased out. Even the home gardener has difficulty with mites, aphids, thrips, cutworms, snails, and countless other small pests which attack his fruits, vegetables, and shrubbery.

The following self-explanatory table was prepared by Dr. L. A. Anderson and appeared in the Shell Chemical Company publication *Scan* in 1972. The data, in pounds or bushels per acre, are given for only six vegetables; however, most other vegetables follow a similar pattern. No chemical is mentioned by name. "Sweet corn" is used as a vegetable, while "corn" is used for corn oil, corn meal, and animal feed.

Citrus growers usually contract with large organizations such as Sunkist to harvest and sell their crops. When an acceptable grapefruit, for example, is purchased by Sunkist, it is marked with the coveted Sunkist stamp. A small citrus grower in southern California had bad luck with his grapefruit crop in

CROP YIELD DATA

	Cabbage	Tomato	Broccoli	Sweet Corn	Corn	Soybean
	lbs/A	lbs/A	lbs/A	lbs/A	bu/A	bu/A
No fertilizer No weed control No insecticide	0	0	0	9,500	88.5	0
Fertilizer No weed control No insecticide	0	0	0	14,000	83.2	0
No fertilizer Mech/hand weeding No insecticide	8,300	8,600	500	14,500	116.0	26.6
Fertilizer Mech/hand weeding No insecticide	7,900	11,000	475	15,550	124.5	41.0
No fertilizer Herbicide Insecticide	15,800	23,500	10,000	18,250	127.4	40.1
Fertilizer Herbicide Insecticide	16,500	30,000	11,750	21,000	143.7	48.0

1970 even though the crop had been grown with the same chemical fertilizers and pesticides that he had always used. For some reason (he believed it was probably inclement weather), his grapefruit were unusually small that year, and his regular buyer refused to buy the crop. After several unsuccessful attempts to sell the crop he took a truckload to a produce stand at the Los Angeles Farmer's Market, where the owner agreed to buy the fruit at a reduced price. Out of curiosity, the farmer returned to the same fruit stand the following day to see if anyone was buying his grapefruit. To his astonishment he saw two bins of grapefruit. One was filled with large, plump fruit over which appeared the sign "Inorganically Grown Grapefruit —2 for 35¢." In the adjacent bin he saw his small orange-size grapefruit under a sign which read, "Organically Grown Grapefruit—45¢ each." He spoke to the proprietor about this and explained that his grapefruit had been grown with the use of pesticides and fertilizers. The proprietor replied, "Hey, I'm glad you're back. Never mind how you grew the damn things, just bring me some more the same size. The things are selling like hotcakes."

A friend in Santa Barbara, California, told me that she could not resist stopping at a strawberry stand in front of which was a large sign reading "Semi-Organic Strawberries." She asked the proprietor what a semi-organic strawberry was. He said, "Well, we put the strawberries in these little plastic baskets and plastic isn't organic."

In an editorial in *California Agriculture,* August, 1971, Dr. Boysie E. Day, Director of the Agricultural Experiment Stations, University of California, wrote the following:

Organic gardening is riding a new wave of ecology-stimulated popularity. Most popular magazines have run articles about the subject and television has covered it extensively. Treatments of the subject range from minor use of unconventional terminology to describe ordinary gardening practices to outpouring of pure nonsense. Television programs proclaim the virtues of organically grown produce and denounce as less nutritious or even harmful, food grown with synthetic or "chemical" plant nutrients and pesticides. On a recent program, the television performer held up a head of "chemically grown" broccoli and announced that it was only 20 percent as nutritious as another seemingly identical head, said to have been grown organically. The organic product was said to taste better and be more healthful than the "synthetic" or "plastic" broccoli.

Green plants are "complete" biochemical factories and require no food in the usual sense, only raw materials. Plants grown in water to which nutrient salts have been added are identical in appearance, taste and food value with plants grown in the richest soil. Nutrients derived from decaying matter are neither better, nor worse, than nutrients from other sources.

Organically grown produce is alleged to be superior because of the natural benefits obtained through organic plant nutrients from composts and manures. This is total nonsense. With the exception of a few parasitic plants such as dodder, higher plants do not utilize organic nutrients. Plants require water, carbon dioxide, and a dozen or so inorganic ions and nothing more. Water, air and a few simple salts constitute a complete nutritional environment for green plants and it is immaterial whether these ingredients are supplied from decaying compost or from a mine or factory.

Dr. Lloyd Henderson, Quality Control Manager of Foremost Dairies, Inc. in San Francisco, California, in an article, "Insecticide Residues in Milk and Dairy Products," in *Residue Reviews,* 1965, has said, "In a discussion of pesticide residues in food products of any kind the inevitability of having some residues, no matter how infinitesimal, must be recognized. Agriculture, as it must be conducted today in an effort to produce food in the quantities and qualities demanded, could not exist without synthetic pesticides."

H. J. Shipman, of The Northwest Processors Association, testifying at the Seattle Hearings in 1969, said: "Much of the reluctance to ban or restrict pesticides arises from the fact that there are too many uses where suitable alternatives do not exist. An all too frequent result of pesticide substitution is the transfer of auxilliary hazard from one group or situation to an entirely different group situation. For example, even though SEVIN [carbaryl, a carbamate insecticide] was tested for several years as a DDT replacement to control corn worms on corn, no one anticipated the degree of honey bee kill that actually resulted. While SEVIN's toxicity to honey bees was accurately predicted the surprise developed with the discovery of the extent to which bees depend upon sweet corn pollen for their winter food supply." [DDT is only mildly toxic to bees.]

At the Seattle Hearings, Dr. John E. Swift, Chemical Coordinator for the Extension Service at the University of California, said that the number of beehives in California has declined yearly since DDT restrictions were enacted. Dr. Swift believes the reason for this is that growers had to use substitutes for DDT that killed bees at a terrible rate. In 1968, for example, 83,000 hives of bees were killed by these substitutes in California. "They estimate that the dollar value of crops that require pollination by bees to be around 300 million dollars. The loss of the beehives has caused a crisis with many growers, not only in increased cost for bees, but in an acute shortage of bees." Although some crops are wind-pollinated, most rely on bees as pollinators.

Clarence Benson, President of the American Honey Producers Association, testified at the Poage Hearings in March 1971: "Three and one-half billion dollars worth of agricultural crops are dependent upon bees for their production, and an additional six billion dollars worth of crops are benefited by bee pollination." Mr. Benson said that when the pesticide SEVIN was applied to cotton crops in the Imperial Valley in California 18,000 bee colonies were destroyed. "As far as beekeepers are concerned," he said, "DDT is a pretty good pesticide."

In his testimony at the Poage Hearings, Dr. George Ware, Chairman of the Department of Entomology at the University of Arizona, stated: "The destruction of our pollinating honey bees is one of the detrimental aspects [of the DDT moratorium]. There were on record in 1960, 108,000 colonies of bees in Arizona managed by 443 beekeepers. Now in 1971, there are estimated to be less than 30,000 colonies managed by 197 beekeepers." Dr. Ware said that the factor responsible for this drastic reduction in bees was the increase of the use of organophosphate and carbamate pesticides on cotton to replace DDT. He concluded, "As an example of recent losses due to pollinating bees, the alfalfa yield (1961 to 1965) was 190 pounds per acre. In 1967 and 1968 the yield was 110 pounds per acre and in 1969 and 1970 the yield was 100 pounds per acre. This is less than half the national average."

According to a 1969 report from the British Advisory Committee on Pesticides, "During the years 1956–1965 when 290 samples of bees (including more than one sample from each alleged poisoning incident) were examined, only two incidents of poisoning by DDT were recorded."

Despite the evidence that DDT has never killed or injured anyone except in the suicides previously mentioned, public pressure against this pesticide has forced many states to place rigid restrictions on its use. Consequently, substitutes have to be used. The Public Health Department of the State of California reported 2,704 cases of systemic poisonings between 1956 and 1968 from the organophosphate pesticides, which are most frequently used as substitutes for DDT. In reference to systemic poisonings, Esther Baginsky, editor of the report, said, "Parathion, as usual, was the most frequently identified of the organophosphates with 106 reports in 1968, followed by Phosdrin, Systox and malathion."

John Noble Wilford told of the tragedy of parathion poisoning in an August 21, 1970, article in the *New York Times*. According to Mr. Wilford, a farmer named Clarence Lee Boyette, from Pink Hill, North Carolina, went to the store to buy a pesticide that would kill the worms that were destroying his tobacco crop. The storekeeper had to sell Mr. Boyette "Big Bad John," the local name for parathion, because DDT had been banned for use on tobacco in that state. The pesticide was "successful." It killed the worms—but it also killed his son. His second son narrowly escaped death. Mr. Boyette said that he had not read the label, but his wife interrupted to say, "If he had read it he wouldn't have knowed those big words."

Mr. Wilford also related the story of seventeen-year-old Jay

Adams, who entered his uncle's tobacco field the day after it had been sprayed with parathion. He became violently ill but after two weeks had recovered enough to return to the field. He died two days later.

Dr. Shirley K. Osterhout, Clinical Director of Duke University's Poison Control Center, was quoted in the Wilford article: "If you ever get poisoned with it, it's advisable not ever to go back again. We've noticed a lot of cases where the victim just piles up daily doses. He seems to be all right, but suddenly he passes out, goes into convulsions or becomes paralyzed."

Dr. Osterhout cautioned that other forms of organic phosphates related to parathion are in many of the insecticides used in homes and family gardens. Dr. Osterhout pointed out a grim irony. She said, "They're worried about that shipload of nerve gas out in the Atlantic and the same thing in a weaker form is in half of the homes and most of the farms."

In November, 1970, the syndicated columnist Jack Anderson wrote about parathion incidents: "In North Carolina alone four persons have died this year from contact with the poison and a total of thirty-two cases have been reported. The Agriculture Department has also investigated parathion poisoning in Alabama, Florida, Indiana, Kentucky, Nevada, New York, Oregon, Pennsylvania, Texas, and Washington. And more than two dozen deaths have been reported from Mexico and Colombia. . . . Dying from parathion is particularly agonizing, as in death from related nerve gases. The victim is first nauseated, then he trembles violently and finally dies in paralysis. Only a speedy shot of the antidote atropine can block the violent effects."

Mr. Anderson also said that the *Agriculture Department Investigative Summary* told of a fourteen-month-old Nevada infant: "A child playing in the yard ate some dirt containing parathion. She later began crying and soon collapsed. Emergency treatment was unable to save her. *Two years* previously her father had treated the yard and patio with parathion. The

parathion placed between the patio bricks did not decompose."
[Emphasis added.]

Furthermore, according to the Anderson article, "In Pennsylvania between 8,000 and 10,000 birds were found dead over
a ten acre area. Analysis revealed high levels of parathion contamination in seed oats. . . . In Washington State an owner
dipped 270 head of cattle in a parathion mixture which resulted
in the eventual loss of 55 head."

Dr. John E. Swift, Chemical Coordinator for the Extension
Service at the University of California, Berkeley, has stated that
cotton is one of the main crops on which he sees no suitable
substitute for DDT. In some earlier cases where DDT was
banned for use in cotton, it was necessary to spray DDT substitutes every three to five days. Nevertheless, the State of California has banned the use of DDT on cotton.

In the summer of 1969, twenty-seven children in Tulare
County, California, were treated for poisoning from organophosphate pesticides. In this cotton growing county, parathion
was substituted for DDT on the recommendation of the California Department of Agriculture, which labeled DDT a
"hard" pesticide. A local doctor, agitated by this, commented,
"Hard on whom, the peasants or the pheasants?"

By the summer of 1970, in Tulare County alone, 15 percent
of the farmworkers were injured by the organophosphate substitutes for DDT.

In their study, "Measurement of the Exposure of Workers to
Pesticides," Dr. William F. Durham and Dr. Homer R. Wolfe
reported, "Numerous illnesses and an occasional death occur
each season in persons occupationally exposed to parathion.
When one considers that a sprayman is exposed to 5.4 percent
of a toxic dose in one hour of spraying or 43 percent in an
eight-hour day, these cases of poisoning are not surprising. In
fact, on the basis of these figures, it would be easily possible for
a sprayman to expose himself to a toxic dose in the course of
two days of work without any accidental exposure. This would

require only that he work 18.5 hours during the two-day period."

It should be noted that spraymen using parathion are urged
to wear protective clothing and respirators. By contrast, no
sprayman or farmer has suffered ill effects or death as a result
of using DDT, regardless of the number of hours involved, and
no protective clothing is needed.

Isn't it time that we took an honest look at the DDT propaganda and considered instead the cold, hard facts of the human
deaths that have resulted from its substitutes? Where are our
perspectives? What are our values?

State Bans:
Trial
and Error

10

Prior to 1972 all states had placed some restrictions on the use of DDT. These varied from limited use for medical purposes only to unlimited use for home gardening with certain restrictions for farm usage. *Most* states had restricted the use of DDT to such a degree that it was virtually banned for agricultural use. The use of DDT was banned on all *federally* owned lands. The restriction placed on the use of DDT by one state in the control of a given insect affected neighboring states. The story of the gypsy moth is a case in point.

The gypsy moth is one of the most destructive insects in the United States. The devastation wrought by this insect is staggering. In March, 1971, Congressman W.R. Poage said that the gypsy moth had been controlled in the past with DDT spray but recently the spray had been banned. As a result, hundreds of thousands of acres of hardwood forest, mostly in the Northeast, had been defoliated by the gypsy moth. He also said that the destruction of these trees would surely have an adverse effect on the wildlife and ecology of the infested area, as well as causing scenic and economic losses. "So the question again arises, should we accept some damage from the use of pesticides or should we let the pest take control and possibly cause even greater damage?"

Dr. Robert White-Stevens of Rutgers University discussed the gypsy moth on a television broadcast on July 14, 1969. He said that in 1968 the gypsy moth denuded over forty thousand acres of hardwoods and softwoods in northern New Jersey, and that wildlife, birds, mammals, and snakes left the ravaged areas in droves because they had no food, no nesting place, and no shade. "There is no valid evidence despite the emotional assertions in the press that any registered pesticide applied correctly and in accordance with approved Federal and local recommendations, has ever invoked massive irreversible bird kills of any species. Even when accidental spills or irresponsible exposures have on occasion occurred, the resulting kills have been insignificant compared to kills elicited by winter wastage, storms, diseases, and such other man-made hazards as skyscrapers, TV towers, and traffic."

According to an article in the *San Jose Mercury,* February 26, 1972, the gypsy moth defoliated more than one-third of the entire forested areas in the State of Connecticut in the summer of 1971. Incidentally, Connecticut is the state that is having problems with the decline in the number of osprey, which build their nests in trees.

The following copy appeared in an article, "Here Comes the Gypsy Moth," in *Reader's Digest,* March, 1972: Last July, at the height of summer's verdant glory, the town of Bristol, Connecticut, stood leafless, its leaves ravaged by an infestation of caterpillars estimated at ten million insects per acre. Hot sun filtered eerily through bare limbs to wither flowers in gardens. Tons of caterpillar feces fell like rain on lawns and sidewalks. People cancelled picnics and cookouts and deserted the streets. Swimming pools were closed. For Bristol and hundreds of other communities in New England, it was one of the worst outbreaks of the gypsy moth in history—twice as bad as 1970's devastating visitation.

"A purely northeastern phenomenon? No longer. During July, 1971, approximately two million acres of suburbs, parks, forests and woodlands from Maine to Cape May, New Jersey, and westward into Pennsylvania and central New York State were defoliated by the gypsy moth."

An example of the devastation caused by the gypsy moth can be seen in the 18,000-acre Newark, New Jersey, watershed. In 1968, 116,693 oak trees were killed by the gypsy moth. The State of New Jersey would not allow the use of DDT, the only known *effective* pesticide that can control the gypsy moth. The authorities chose instead to use "biological control" in the form of the *chemical* sex attractant, Disparlure. By October, 1971, the total of dead oak trees in the Newark watershed had risen to 1,055,820. This figure constitutes 58.5 percent of the oak trees in that area. Other trees in the Newark watershed under attack by the gypsy moth include hemlocks, red maples, hickories, birches, sugar maples, and ash.

In a telephone conservation with Stanley A. Hall, of the United States Department of Agriculture Research Station, Beltsville, Maryland, on April 11, 1972, I was told that all of the New England states were badly infested by the gypsy moth and that its destructive trail had moved northward into the province of New Brunswick and as far south as Virginia, Pennsylvania and New York. Mr. Hall was asked what was being done. He said that they were very optimistic that the Disparlure would work in Virginia because at this point Virginia had only a light infestation. I asked him why these states didn't use DDT and he said that it could not be used because it was not allowed. Mr. Hall also told me that the spruce budworm is back at work in Maine and the forests of New Brunswick.

According to the *Reader's Digest* report: "So far, the gypsy moth has cost the U.S. government and cooperating states over 100 million dollars in research and eradication funds, and before its devastion is controlled this leaf-eating hitchhiker may prove to be one of the most expensive insects in the world."

Pines, hemlocks, and spruce die after a single defoliation by the gypsy moth. Most hardwood species produce new leaves four to six weeks after defoliation. After a second defoliation most hardwoods die.

The gypsy moth larvae hatch from eggs laid the previous summer. In late April and early May the tiny caterpillars, only one-eighth of an inch long, emerge from the egg masses, each of which may contain from 100 to 1,000 eggs. The little caterpillars climb to the top of a tree and begin to feed immediately. If there are too many caterpillars competing for food, the caterpillar spins lengths of silk which enabled it to be carried by the wind to other trees; hence the term "gypsy." A storm in the spring of 1971 carried gypsy moths from Connecticut to Long Island, New York, over distances of fourteen to twenty miles.

The caterpillars molt from four to six times before going into the pupal cases from which they emerge as adults. During its final days as a caterpillar, this creature is capable of eating five to ten leaves during its nightly feeding period. A badly infested tree can have as many as thirty thousand caterpillars feeding upon it at a single time.

The adult moths emerge from the pupal cases in mid-July. The female emits a powerful lure which can attract a male from a distance as great as a quarter of a mile. It is this lure that has been duplicated chemically in the form of Disparlure and is being tested experimentally.

The female is so heavily laden with eggs that she cannot fly. She is not particular about where she lays her eggs, which have been found under lawn furniture, cars, and trailers. The news media have asked travelers to check beneath their cars and trailers for the egg masses, to be sure that they do not transport the gypsy moth to even more states. It is ridiculous to assume that people, many of whom are not aware of the gypsy moth problem, and who would not recognize the egg masses if they *did* look, would crawl under their vehicles every morning to search for gypsy moth eggs. Entomologists in all areas are concerned that this "gypsy" will sweep its way across the country.

It is necessary to distinguish between insect control and insect eradication. Control is the reduction of an insect population to reasonable levels. Eradication is the complete elimination of an insect from a geographical area. At this writing, entomologists are fighting for *control* of the gypsy moth. Most entomologists believe that it is presently impossible to *eradicate* the gypsy moth from this continent.

When a forest dies, there is an increased danger from fire and erosion. Stream flows are altered and wildlife is drastically affected. The destruction caused by the gypsy moth was described in an article in *Barron's,* July 6, 1970, under the headline "Ravaged Summer."

" 'We are in a state of emergency,' cried one resident of Shirley, Long Island. 'Our children cannot go out. Our pools are finished for the summer. It's a question of survival—the caterpillars or us.' "

The article continued, "A double invasion by the gypsy moth and canker-worms has stripped trees, driven away picknickers [sic] and campers, and caused widespread concern in upper Westchester County and Western Connecticut."

The article was specific about the size of the problem and its causes. "Of the 25,000 acres severely infested and in danger of defoliation, the State of New York, against the bitter opposition of conservationists, has sprayed only 14,000 acres in areas of low population density; not with DDT, which it shuns, but with SEVIN, a less effective, short-lived insecticide. In the northern part of New Jersey, which stopped the aerial spraying of DDT over half-a-decade ago, the gypsy moth for the second straight year has damaged thousands of acres of woodland, and, to the consternation of local inhabitants, caused a mass migration of copperheads and rattlesnakes to lower, shadier ground."

It was against the Suffolk County Mosquito Commission that Victor J. Yannacone, Jr., former attorney for the Environmental Defense Fund, filed suit on behalf of his wife and "all other people of Suffolk County" because the Commission's dumping of DDT was alleged to be the cause of a fish kill in a lake near Mrs. Yannacone's childhood home.

Robert J. Sullivan, from Pennsylvania, testified in behalf of the National Christmas Tree Growers' Association at the Poage Hearings in March 1971:

The gypsy moth is epidemic in the northeastern part of this country. I am not being an alarmist at all when I say that it threatens the entire forestry chain of the Appalachian Belt from New England to Florida. It has become a very severe threat. Let me read just a word from the meeting of the National Gypsy Moth Advisory Council of November 17, 1969. This is picked at random.

"The moth is now out of control. Eradication is now impossible. Eventual control is the best that can be expected. It will take five years for a quantitative analysis of the situation necessary for an intelligent control program. It will take at least 10 years for the establishment of integrated control and possibly 100 years."

Now, this has occurred because DDT has been stopped in the control, in the eradication of the gypsy moth. The gypsy moth started about 1860 and at first was controlled in New England by arsenates which have been abandoned because they are toxic. DDT was then taken up and the pest became controllable and was practically eliminated. Then, because of public opinion, which I must say was mostly uninformed, DDT was abandoned and the gypsy moth has gone ahead. The result has been just what I have read: It is out of control.

The national council consists mostly of technical men, of which I am not one. They are entomologists and silviculturists. Most are State employees. They are public servants. I said to them, why not return to DDT? It has never been proven that DDT does any harm, you know, and we know the lots of good that it does do.

These men say, no, we cannot do that.

I say why?

They say because of public opinion. They are public servants. I do not blame them. I would be in the same boat if I were a public servant.

So what do they propose instead? They would propose, and Senator Byrd of West Virginia has offered a bill in Congress for this purpose, that a 5-year program be started which will cost $3,725,000 for the study of this situation and that after that study has been made, a start be made on the control—not the eradication—of the pest. Now that $3,725,000 would be used up before it was finished. We all know that. That $3 million would be saved if someone would have the guts to get up and say, let's use DDT. But nobody will do it.

I, as a Christmas tree grower, am in a quarantined zone. I cannot ship Christmas trees without complete inspection of my product by State inspectors to make sure that I am clean. If the gypsy moth should get into my plantation, I could lose it all in one season. Deciduous trees usually survive 2 or 3 years of defoliation, which is what the larvae of the gypsy moth does. Coniferous trees would last only one season.

As previously stated, the United States Department of Agriculture has reported that the gypsy moth has gained a foothold in Pennsylvania. Mr. Sullivan's plantation is now in jeopardy.

In an article in the *Portland Oregonian,* December 28, 1969, Dr. J. Gordon Edwards wrote: "The condition of Orgeon's forests is of great concern to Oregonians, and DDT plays a most important role there. When the Douglas Fir Tussock Moths threatened thousands of acres near Burns, Oregon, Forest Service officials used DDT to control them. An excellent 'Surveillance Report' (1968) documented the lack of ill effects on fish, wildlife, and cattle. It was stated that: 'DDT still remains the only effective insecticide for control of the Douglas Fir Tussock Moth.' (Ultimate losses prevented by the spray program were estimated to be more than $16 million.) Another excellent report in 1969 dealt with the highly successful Willipa hemlock-looper control program, and proved that DDT did not adversely affect non-target organisms in the forests, streams or bays."

Oregon banned DDT for use on these pests in 1971. The State of Washington had also banned DDT for virtually all uses. The importance of forests to the economy of both Oregon and

Washington cannot be emphasized too strongly. In Washington, one-fifth of the total "value added by manufacturing" is attributable to forest products. Oregon produces more than one-fifth of the lumber used annually in the United States and produces more than 65 percent of the nation's plywood and 40 percent of its hardboard and fiberboard.

In an April 9, 1972 telephone interview with horticulturist H. J. Conklin of Cashmere, Washington (in the foothills of the Cascade Mountains), Mr. Conklin said, "The once beautiful forests within five miles of my home are dead. Last summer [1971] it looked like a giant fire had swept the area. We pleaded with the state to use DDT. They refused. If they're going to sit on their hands and watch our forests die from insect defoliation, why do they bother spending millions of dollars to control a forest fire? The end results are the same. Naturally, I'm not proposing that they abandon fire control, but it makes about as much sense. To waste time trying one pesticide after another while forests die and wildlife is depleted makes about as much sense as attacking a forest fire with a squirt gun."

Widespread forest spraying programs using DDT to control forest insects have had surprisingly little effect on birds, fish, and mammals. Even in rare cases where non-target organisms have been affected, the population levels quickly returned to normal. When a coniferous forest is destroyed, it will take from thirty to forty years for pulpwood production to resume.

Maine had banned the use of DDT; however, in 1967 when the spruce budworm threatened to destroy her forests, the state hastily reversed its decision.

Once again, where are our values? DDT has proved itself to be effective in the control of the gypsy moth. It is one thing to be optimistic about biological control, but quite another to be downright foolhardy. How long are we to stand by and watch the gypsy moth defoliate our forests, and hence threaten a large part of our economy and our wildlife? Certainly, in the case of the gypsy moth, there is concrete proof that the *lack* of DDT is raising havoc. We can all be hopeful, as was Rachel Carson, that biological control is just around the corner. Until we reach that corner, it is only logical to use DDT to protect our forests, forest industries, and our wildlife.

The farmer, like the forester, has many chemical pesticides available to him. He may use them *if* they will control the damaging insects on that crop and *if* they are economical enough to leave him a reasonable profit. The cost is not only in the price of the chemical, but also in labor and equipment needed for repeated applications. A safe, inexpensive, long-lasting pesticide can spell the difference between a marketable crop sold at a profit or a loss. Since DDT has been banned several applications of an alternate pesticide are necessary to control the same pests. The cost has been passed on to the consumer.

Arden W. Jacklin, of the Jacklin Seed Company, Inc., Dishman, Washington, testified at the Seattle Hearings on October 16, 1969. Mr. Jacklin's company raises extensive amounts of lawn seed for commercial sales. Mr. Jacklin said, "Our farms, with the surface soil well charged with DDT and with annual applications of granules of DDT, now have a good level of bird as well as small animal population."

He continued, "We know our farms have good wildlife populations. We hunt on them ourselves. We have a 15 year history of application of DDT to our soils and our observations are that our wildlife population has increased rather than decreased."

Mr. Jacklin said that his company is quite successful in producing seed of four varieties of perennial rye grass on soil fortified with DDT. His land had been systematically treated with DDT, averaging three pounds per acre per year.

Alfred L. Long, chief fieldman for the Inland Empire Pea Growers Association, in his testimony at the Seattle Hearings, said that when the pea crop is starting to bloom, the pea weevil moves into the crop from the sides of the fields. Its only purpose is to lay eggs on the pea pod which is developing from the blossom. The larva hatches and in a few days burrows through the wall of the pod into the pea itself. Some larvae die leaving "pinhole" weevil damage, but others eat much of the pea itself causing light peas.

Mr. Long also said, "By dusting or spraying with DDT as soon as the infestation of the field is observed, it is normally possible to attain control with a single application of one pound of actual DDT per acre on about 25 percent to 40 percent of the acres in the field. By bordering the field with a strip about 100 to 150 feet wide, and also up the draws, the weevil in the primary infestation are killed. Because of the residual potency of the DDT, an effective barrier is provided to guard against re-infestation through the blooming period of the peas. The cost of the DDT and the application are about $2.10 to $2.25 per acre.

"The alternatives are Malathion or Parathion. Malathion application costs about $3.50 to $3.80 per acre and Parathion costs about $2.80 to $3.00 per acre for a single application. However because they are only effective for 24 to 48 hours after application, no barrier to re-infestation is provided. Two, three and even, in severe condition, four applications may be necessary to protect the peas through the two or three week blooming period. This, of course, can greatly increase the cost of production.

". . . . Even with prompt use of DDT, 2 to 3 percent weevil damaged peas are fairly normal, and 8 to 10 percent is not unusual . . . Since dry peas are such a low income producing crop, our growers can stand neither the increased production costs nor the potential loss in weight and grade to be expected from the alternatives to DDT. . . . Until an effective, low cost, alternative to DDT is available, we believe a complete ban on DDT would seriously curtail the production and exporting of dry peas. With no crop substitute for peas, the financial burden on our growers could be disastrous."

At the Seattle Hearings in October, 1969, Don Mock, president of the Pacific Northwest Pesticide Applicators Association, Washington Chapter, testified, "Without DDT, all the holly in the Puget Sound Region would have to be sprayed within a five day period in order to control the holly-leaf miner.

"Without DDT, the cost of control of leaf miners on lilacs, privet, birch, madrona and many other plants would be almost prohibitive. Without DDT, weevil and beetle damage in our ornamental plantings would be devastating.

"Without DDT, control of our perennial pest, the tent caterpillar, would be difficult if not economically impossible. In a season like the spring of 1968, when caterpillars hatched over a span of 30 days, the use of a long residual pesticide was imperative."

Mr. Keith Sime, sales manager for Miller Products, a chemical company in Portland, Oregon, in his testimony at the Senate Hearings, said, "DDT has more clearances and accepted use for insect control on food, fiber ornamentals, and industrial usage than any other pesticide on the market. It has been, for many years, the wheel-horse, the bread and butter, or the all-around use pesticide."

Mr. Sime also said, "In some instances DDT cannot be replaced with any other chemical to give as effective economic control. This is true with the mint flea beetle and with loopers and other caterpillars. Any substitute now cleared has to be used much more frequently. Due to the attractiveness of mint for bee pollinators, where substitutes for DDT are used, there is considerable loss of pollinators. We have no ax to grind. A state ban on DDT would mean more business for us as an industry."

Mr. Sime said that prices would skyrocket and the consumer would pay the bill. They have, and we are.

H. J. Conklin, the horticulturalist with whom I spoke in April, 1972, also testified at the Seattle Hearings: "A quick glance back into the history of DDT, which came to the fruit growers of our area [Wenatchee, Washington] in the mid-1940s, tells us that its importance to the fruit grower in pest control was comparable to the medical world and development of penicillin and the many lives it has saved. DDT brought to an end the devastating and staggering losses of hundreds of thousands of dollars worth of fruit that were being ruined annually by the codling moth, which entered the apple in the larval or worm stage, rendering it a cull and unfit for sale to the consumer.

"It was not uncommon for a grower of apples prior to the use of DDT to sustain a loss of 15 to 30 percent of his fruit even though he had sprayed his orchard 10 to 12 times during the growing season with lead arsenate, the most commonly used material in the 1930s and 1940s.

"The advent of DDT and its general usage brought this loss of fruit, caused by worms, and thusly the loss of dollars to the economy, to a near nothing figure. It might be said that DDT could be credited with saving the apple, pear, and stone fruit industry of the State of Washington and gave it the life to go on and become the important business that it is today.

"Other alternative compounds which could be used for the control of these insects are economically impractical or less desirable than DDT. Less desirable because less is known about their efficacy on the control of these pests. Also, because these other materials have been used less frequently and for a shorter period of time, less is known about their potential deleterious effects on soils, beneficial insects and to say the least, on man, domestic animals, fish and wildlife."

In my telephone interview with Mr. Conklin, he said that the State of Washington had banned the use of DDT on fruits. "I cannot say that we would not survive without DDT up here, but I can say that we could do a far better job with it. Our backs are against the wall as profits drop and prices rise."

The state of Washington spent a fortune on DDT Hearings in October of 1969, and as a result of the hearings placed rigid regulations on the use of DDT. However, after the state had suffered heavy economic losses to apple, pear and mint harvests, the State Pesticide Control Board met on April 11, 1971 to reverse its decision and allow DDT to be used against the cutworm on these crops.

Mr. Reuben S. Gervais, manager of the Thompson-Hayward Chemical Company of Yakima, Washington, capsulized a major problem regarding pesticides in his testimony at the Seattle Hearings: "Because of the length of time and huge costs involved there have been relatively few new insecticides that have been developed in the past ten years. It is an alarming fact that our vast food surpluses have dwindled as more and more of our farm output is exported to hungry nations and that our huge productive capacity of U.S. agriculture will be strained to just keep up with the food demand. Without the use of pesticides it will be virtually impossible to produce more food on less available land."

In his article in the *Portland Oregonian,* cited earlier, Dr. J. Gordon Edwards said: "The President recently launched a campaign to 'end hunger and malnutrition in the United States.' That program is doomed unless DDT *and* its allies remain available for use against agricultural pests. Food prices will soar if DDT is banned for substitute pesticides must be applied four to fifteen times more often and each application costs two to five times as much as a DDT treatment. At this writing, Dr. Edwards' 1969 prediction has become a reality.

"In October, 1969, 8,000 Washington orchardists petitioned the Department of Agriculture not to ban DDT in that state and California cotton growers found that production costs increased seven-fold when substitutes for DDT were used."

DDT is ineffective against many insects, particularly "domestic" insects such as the housefly. Other insects can be controlled by substitute pesticides if we are willing to pay the price in human misery and in dollars. There still remain many insects that can be controlled only by DDT. Money is desperately needed for extensive and *realistic* research in this area. A safe, economical substitute for DDT must be found because the development of insect resistance to *any* pesticide must be considered inevitable.

Until the time that reasonable research *does* provide mankind with such a substitute, DDT must be used. Otherwise children will continue to die from parathion poisoning, prices will continue to soar, labor costs will rise because of the need for repeated pesticide applications, more forests will be defoliated, and our wildlife *will* be endangered. To ban DDT and wait optimistically for biological control, which obviously is not yet effective, is to "feed" on false hope which may well result in malnutrition.

One
Scare After
Another

11

Pesticides, chemical fertilizers, and detergents do not stand alone in the whirlpool of confusion and misinformation brought about by exaggerated and unrealistic experimentation. A most confusing "set" of laws has placed our nation at the mercy of regulations which, though they appear meritorious, are, in fact, detrimental.

Section 402 of the Food, Drug, and Cosmetic Act reads: "A food shall be deemed to be adulterated— (a) (I) If it bears or contains any poisonous or deleterious substance which may render it injurious to health; but in case the substance is not an added substance [i.e., a natural one] such food shall not be considered adulterated under this clause if the quantity of such substance in such food does not ordinarily render it injurious to health . . ."

This law, though doubtless enacted with the best of intentions, provided a double standard and has been exploited to such a degree that countless people fear all things not naturally present in foods. A close look at this law shows that if a poisonous or deleterious substance is *naturally* present in a given food, that food is deemed to be safe. If the *same* substance is *added* to a given food which does not *naturally* contain this given substance, it is, by law, not safe.

This law states, " . . . does not ordinarily render it injurious to health . . . " Because they did not qualify the word *ordinarily* and because they neglected to state injurious to the health of *human beings,* it became possible under the law to exploit unrealistic experimentation with test animals.

Further complicating this legal double talk is a clause in Section 409 of the Food Additive Amendment (Pub. Law 85–929) which was swept into law on a tide of emotionalism in 1958. Shortly after his wife had died of cancer, Representative James J. Delaney (New York) presented the following amendment (the Delaney Clause) before Congress: "That no additive shall be deemed to be safe if it is found to induce cancer when ingested by man or animal, or if it is found, after tests which are appropriate for the evaluation of the safety of food additives, to induce cancer in man or animal . . . "

It is difficult to think of a more humanitarian objective than is implied by this clause. It is even more difficult to comply with it, because as previously stated, innumerable substances can be shown to cause cancer in animals, depending upon the dosage, the means of application or administration, the frequency of dosage, and the susceptibility of the test animal.

Dr. Richard L. Hall, Vice President of Research and Development for McCormick and Company, prepared what he termed a "reasonably elaborate and attractive but not at all exotic dinner menu":

THE MENU

Radishes, carrot sticks, celery

Smoked salmon

Seafood in patty shell

Glazed ham

Lima beans　　　　　　　　　　　Creamed spinach
Baked potato with　　　　　　　　Cauliflower with
　　sour cream and bacon　　　　　　　Hollandaise sauce
　　　　　　　　Candied turnips

　　　　Mixed green salad—Roquefort dressing
　　　　　　Rolls　　　　　　　　Butter

　　　　Camembert cheese and crackers
　　　　Compote of assorted fruits
　　(bananas, pineapples, strawberries, peaches, pears)
　　　　Cashews, almonds, peanuts

　　　　Cocoa, coffee, cola, milk and tea

　　　　　　Wines and liqueurs
　　　　Creme de menthe　　　　Cointreau
　　　　　　　Cognac

I have prepared a list which shows each food on Dr. Hall's
menu, and have indicated the "harmful" ingredients which are
naturally present in each. If these foods were to be judged by
the standards which are set for food additives, every food would
have to be classified as "deleterious" even though the "harmful" contents are minuscule. Note that I have not included the
crackers. They are "safe" provided they are unsalted. Salt is
toxic—again dependent upon the amount ingested.

"NATURAL FOOD"	"HARMFUL" INGREDIENTS NATURALLY PRESENT
radishes, carrots, celery, potatoes	cholinesterase inhibitors which interfere with the transmission of nerve impulses
almonds, lima beans	hydrogen cyanide precursors

"NATURAL FOOD"	"HARMFUL" INGREDIENTS NATURALLY PRESENT
spinach, cashews, tea almonds, cocoa	oxalic acid and oxalates, both toxic
wines and liqueurs	alcohols, some very toxic
peanuts, turnips, cauliflower, peaches, pears, strawberries, spinach and carrots	all contain goiter producing agents (e.g., thiocyanate)
bananas, pineapple, cheese (particularly Camembert), wine	pressor amines which raise blood pressure
egg yolk, butter, milk	vitamins D and A, both stored in body fat— also contain cholesterol
seafood	zinc, arsenic, and mercury
ham and bacon	cholesterol, nitrites or nitrates
smoked salmon	polynuclear aromatic hydrocarbons—dietary carcinogens
rolls (yeast)	ricket promoting factor— amino acid imbalance
patty-shell (shortening)	cholesterol
cola, tea, coffee	caffeine, a stimulant and mutagen
cocoa	theobromine, a toxicant and stimulant

A small verse with a large message was written by Congressman Jamie L. Whitten (Mississippi), Chairman of the Appropriations Subcommittee, Agriculture, Environmental and Consumer Protection, of the House of Representatives:

> So let your life be ordered
> By each documented fact,
> And die of malnutrition
> With arteries intact.

The Delaney Clause is not highly regarded by most responsible scientists but it has become the bread-and-butter of those who use it to suit their own purposes.

CRANBERRIES   The first widely publicized incident involving experimentation which misused the Delaney Clause was the "cranberry scare."

One week before Thanksgiving, in 1959, Dr. Arthur Flemming, then Secretary of the Department of Health, Education, and Welfare, alarmed the public by his announcement that cranberries contained traces of a weed killer, aminotriazole, which *could cause* cancer.

The American Cyanamid Company had tested aminotriazole in rats and found that very high levels of this chemical in the diets of these laboratory animals cause some of them to develop thyroid cancer. To reach the same level of concentration, a human being would have to eat several hundred pounds of cranberries *every day*. Anti-thyroid compounds which act in a manner similar to aminotriazole are present naturally in cabbage, turnips and many other vegetables (see page 152).

Dr. Virgil H. Freed of Oregon State University, Corvallis, tested many samples of the 1959 cranberry crop for the Department of Agriculture and stated that the highest concentration of aminotriazole that he was able to find in *any* sample was only 0.05 ppm. Dr. Freed said, "You could eat these cranberries every day of the year and not suffer any adverse effects."

On November 9, 1959, the president of Ocean Spray Cranberries, Inc. appealed to Dr. Flemming to delay the press release for forty-eight hours to enable the cranberry industry to investigate the problem and confer with him in order to avert a catastrophe to the industry and growers. He urged the Secretary not to issue a press release unless he felt that there was a clear and imminent danger to public health. The appeal was denied, for reasons the Secretary stated in a press release issued later the same day. Dr. Flemming, in the release, urged that no further sales be made of cranberries and cranberry products grown in Oregon and Washington in 1958 and 1959. He advised that sales be suspended until the cranberry industry developed a workable plan to separate the "contaminated berries" from the uncontaminated ones.

The Secretary also stated within the press release that although there *may be* a minute proportion of a carcinogen which is safe in foods consumed by human beings, the FDA scientists did not know whether this was true or how to establish a safe tolerance for human beings.

The president of Ocean Spray stated that he felt that Secretary Flemming should not have made the press announcement because as a result all cans of processed cranberries and all packages of fresh cranberries in markets across the country were placed under suspicion during the crucial two-month holiday period when the industry does 65 percent of its annual business.

By January 1960, 996,000 barrels of cranberries valued at 21,500,000 dollars had become "surplus." Thousands of tons of cranberries were dumped and economic losses to the cranberry farmers were disastrous.

The president of Ocean Spray said in 1964 that it had taken five years for the company to regain its pre-1959 price level. During the seven years prior to 1959, 35,000,000 dollars had been spent to advertise and promote cranberry sales. In an attempt to help recoup their financial losses, the now popular cranberry juice was developed. The cranberry fiasco resulted from inadequate laboratory experimentation on test animals and misinterpretation by scientifically naive FDA administrators.

CYCLAMATES   The cranberry "ban" resulted in heavy financial losses for hundreds of farmers. The total ban on cyclamates has affected *millions* of consumers.

In an editorial in *Barron's,* February 7, 1972, Robert M. Bleiberg commented, "For its money—if the recent past be any guide—the American people may get more than it bargained for. In late 1969, on the basis of experiments involving only twelve rats and after a hurried meeting of the scientific powers-that-be, the Department of Health, Education and Welfare, at FDA's behest, banned 'forthwith' the use of cyclamates in the production of foods and beverages." Mr. Bleiberg reported that Dr. Jesse Steinfeld, Surgeon General of the United States, had said, "There is absolutely no evidence to demonstrate in any way that the use of cyclamates has caused cancer in man."

The experiment which prompted the FDA to ban cyclamates has yet to be duplicated. According to an article in *Drug Trade News* (a private concern not connected with the FDA), January 10, 1972: "Food and Drug Research Laboratories, which had found bladder cancer in rats on cyclamate (the work that led to the removal of cyclamate from the market) is having trouble duplicating its findings. FDRL told this to the Calorie Control Council but cautioned that results were not all in and that findings must be considered preliminary. They have examined one slide each on more than 200 rats given up to 150 mg/kg cyclamate daily for two years."

This article reported that an Abbott Laboratory pharmacologist told the Calorie Control Council that new laboratory studies have failed to duplicate the original findings of Dr. Marvin Legator of the FDA. The article concluded, "The FDA was under a lot of pressure two years ago, on the cyclamates, but perhaps the agency should have sat still until the studies were repeated."

According to a report prepared by John D. Furrer of the University of Nebraska, "The restriction of cyclamates was based on tests which showed that high levels of this ingredient induced cancer in rats. However, a human being would have to drink between three hundred and six hundred bottles of cyclamated beverages, depending upon the brand, *at one time* in order to equate the consumption by the rats."

Obesity from the overconsumption of sugars, starches, and fats is a leading factor in causing premature death in this country. Excessive sugar contributes heavily to dental cavities in younger children, and excessive sugar sometimes interferes with the treatment and control of diabetes, and indirectly contributes to heart disease.

Cyclamated drinks allowed mothers to give their children soft drinks without fear of spoiling appetites or causing tooth decay. They allowed people with weight problems to enjoy food and drinks that they previously could not consume. Tons of cyclamates were ingested by millions of people over a twenty-year period with no recorded rise in the number of cases of bladder cancer.

Howard J. Sanders, in an article, "Food Additive Makers Face Intensified Attack," *Chemical and Engineering News,* July 17, 1971, said, "Many people in the food additives field believe that the current widespread public concern about food additives was touched off by the Government's 1969 announcement that the non-nutritive sweetener, cyclamate, was no longer to be classified in the GRAS category (Generally Recognized As Safe)." The article also said, "FDA was required to prohibit the use of cyclamate in food because the Delaney clause of the Food Additive Amendment forbids the use in foods of any additive known to produce cancer when ingested in any amount by man or animal."

Mr. Sanders concluded, "The furor over cyclamate, which

dragged on for almost a year after the compound was removed from the GRAS list, had repercussions throughout the food additive industry. If cyclamate could produce cancer in animals and possibly in man, the public reasoned, might not other food additives 'generally recognized as safe' also be threats to human health? Was the FDA too lax in regulating the chemicals (sometimes referred to damningly as 'those noxious chemicals') used in foods? In the minds of many people these questions still persist."

The cyclamates were banned on the basis of unrealistic laboratory experimentation on *twelve* rats. At this writing, despite evidence that cyclamates are not harmful, but in fact can be helpful, neither the FDA nor HEW has reversed its decision.

SACCHARIN   Saccharin has been in wide use for over eighty years. Since the ban of cyclamate, saccharin is the only non-nutritive sweetener left for use in the United States in low calorie foods and drinks.

According to Robert M. Bleiberg in a *Barron's* editorial February 7, 1972, "None the wiser, FDA on the basis of experiments with rats, conducted by the Wisconsin Alumni Research Foundation [WARF] (supported by the sugar industry), has begun to look askance at saccharin." These rats were fed enormous amounts of saccharin over a two-year period. In order to keep up with the WARF rats, a human being would have to consume 875 bottles of diet soft drinks per day. Mr. Bleiberg said that only a few of the experimental rats showed signs of bladder tumor. The accusation that the sugar industry supported the saccharin research *was made by Barron's*. The author's attempts to verify this with WARF were fruitless.

Over the years, there have been a few scientists who have questioned the safety of saccharin. In his July, 1971, article in *Chemical and Engineering News,* Howard Sanders wrote: "As recently as April, Dr. George T. Bryan of the University of Wisconsin told an American Cancer Society press seminar that saccharin is a potential carcinogen in man. The test method he used indicates, he says, that saccharin can cause bladder cancer in mice. Other scientists, however, doubt the validity of Dr. Bryan's test procedure which involves the implantation of a cholesterol pellet containing 20 percent saccharin in the bladder of mice.

". . . . In July 1970, a National Academy of Science/National Research Council committee told FDA that 'the present and projected use of saccharin in the U.S. does not pose a hazard' to human health.

"Last month, however, FDA announced plans to remove saccharin from the GRAS list in the near future and to limit its use to specified maximum concentrations in various types of foods, such as soft drinks, fruit juice and processed foods. Since the present concentrations of saccharin in foods are much lower than these limits, no changes in existing foods will be required."

Mr. Sanders explained that the objective of this FDA action is to limit the intake of saccharin to one gram per day for an average adult. Few people ingest more than 0.2 gram per day.

Scare headlines and sensationalism regarding food additives which are in common use do little to ease the public's apprehension. An example of this was also given in Mr. Sander's article: "Contrary to a recent headline in the *Washington Post* ('Saccharin Off Safe List'), FDA does not now regard saccharin as unsafe." Where are the headlines announcing that the National Academy of Science considers saccharin to be safe?

MONOSODIUM GLUTAMATE   Following the ban on cyclamates, monosodium glutamate came under attack in the flurry of concern and confusion regarding food additives. Many people became confused because of the similarity between the two words *cyclamate* and *glutamate*.

Monosodium glutamate (MSG) is a salt which differs from regular table salt (NaCl) in that it is a salt of glutamic acid, an amino acid. Amino acids are the "building blocks" used by nature to produce protein. Glutamate is found *naturally* in almost every protein food we eat—such as beefsteak, cheese, tomatoes, mushrooms, and even mother's milk. There is no difference between the form of (natural) MSG and the MSG which is added to foods by cooks and canneries. The total quantity of glutamate present in a high protein food can be as high as 43 percent.

MSG has been used for hundreds of years, especially in the Orient, to enhance the flavor of foods, particularly those containing protein.

The monosodium glutamate controversy began quietly, almost innocently, when Dr. Ho Man Kwok wrote a letter to the editors of the *New England Journal of Medicine* in early April, 1968. Dr. Kwok reported certain symptoms he experienced whenever he ate at a Chinese restaurant. These symptoms later became known as the "Chinese Restaurant Syndrome."

Two doctors from the Albert Einstein School of Medicine in New York reported to the same medical journal that informal preliminary experiments *suggested* that the cause of "The Chinese Restaurant Syndrome" was MSG.

Though this medical journal considered the whole thing to be a "put on," the news media went considerably further in their treatment of the subject. On the very day that the medical journal referred to the symptoms as "The Martini Syndrome," the *Wall Street Journal* announced "MSG Zeroed in on by Some Scientists as the Cause of the Mysterious Malady."

Three days later a *Washington Post* article flatly stated, "The cause is the flavor enhancer, MSG." Three days after that a reporter for WCBS-TV News in New York announced, "A very small quantity of MSG will cause the illness."

On KSD-TV (St. Louis, Missouri), September 22, 1970, Dr. John Olney of Washington University stated, "There is no doubt that MSG is dangerous for the infant nervous system." Dr. Olney based this conclusion on his experimentation with *newborn* baby mice and *a* monkey. The doctor had discovered brain damage in a baby monkey into which MSG had been *injected* subcutaneously in massive amounts. MSG is *ingested* by, not injected into, human beings. Dr. Olney also reported finding brain damage in *newborn* baby mice into whose stomachs MSG was *pumped* directly by means of a *tube*. In an editorial in *Food Technology,* January, 1970, Karl Herz wrote that the alleged harm from MSG had not stood up under the test of scientific examination and reproducibility: "The 'evidence' is based on administration of unusually high dose rates by unusual methods (for example, injecting under the skin), which in the words of eminently qualified scientists, does not have 'any relevance to the question of safety of MSG as a food seasoning agent.' "

The National Academy of Science/National Research Council reaffirmed that baby mice and baby rats were inappropriate animals in the MSG study because their brains, at birth, are poorly developed.

On WLS Radio in Chicago, September 26, 1970, Ralph Nader said, "MSG has no value in baby food and might be harmful."

In an article in *Food/Chemical News,* September 28, 1970, Mr. Nader wrote, "Even if MSG is proven to be safe, it shouldn't be permitted in baby food."

By October 20, 1970, Mr. Nader was in Dallas, where he announced on WFAA-TV, "Other chemical additives used in food demand more attention than cyclamates, one is MSG."

Finally the manufacturers of baby food announced that they were *voluntarily* withholding MSG from their products because of public fears, despite the fact that they considered it safe for public use. There are some individuals who are allergic to MSG, strawberries, penicillin, and even to aspirin.

The National Academy of Science/National Research Council, in a report to the FDA, have stated that their findings show no deleterious effects to human beings from MSG except to those persons who might be sensitive to the substance. The report also stated that the flavor-enhancing property of monosodium glutamate was considered to be *beneficial* and no restrictions had been placed upon the product. I have not seen a single headline announcing that MSG is considered not only safe but also beneficial.

HEXACHLOROPHENE (pHisoHex) Hexachlorophene (HCP), a widely used anti-bacterial agent, has been an ingredient in a wide array of products ranging from deodorant soaps and sprays, toothpastes, mouthwashes and throat lozenges to furnace filters and is present in germicidal cleansers such as pHisoHex.

The story of pHisoHex once again points out the danger of widespread publicity given to unrealistic laboratory experimentation. In this case an overzealous press has succeeded not only in confusing the public but also in casting a shadow of doubt on the credibility of those scientists who have been and are continuing to use orthodox scientific procedure in order to get to the root of one of the thorniest medical issues in many years.

Initially the FDA played a game of cat and mouse regarding the efficacy of HCP in preventing and controlling staphylococcal colonization and disease. In their press releases they aired these questions in public. The FDA ignored the protests of medical specialists in the field of newborn care who urged the continued use of HCP to prevent outbreaks of staph infection in their nurseries, and instead took a public stand against the germicidal cleanser on the basis of unrealistic animal experimentation.

After experiments that showed brain damage in rats and monkeys who had *ingested* HCP, and detectable blood levels in infants that had been washed with HCP, but with no evidence of human infant toxicity, the FDA issued a series of warnings. At first the FDA did not *ban* pHisoHex. The agency publicly announced that it was unsafe and advised against its use. This announcement placed doctors and hospitals across the nation in double jeopardy.

If pHisoHex had been used and the infant had developed

identifiable toxicity, both the doctor and the hospital could have been sued for ignoring the *advice* of the FDA. If pHiso-Hex had not been used and the infant had contracted staphylococcal infection, the doctor and the hospital could have been sued for not using every means at their disposal to protect the infant.

Later the agency modified its directives to recommend that pHisoHex be used only in cases of emergency. This meant that instead of permitting the use of pHisoHex as a preventative germicide the FDA sanctioned its use only *after* an outbreak of staph developed in a nursery. It takes between twenty-four and forty-eight hours to complete a culture to identify staph; however, sufficient evidence is usually available at the end of a twenty-four-hour period to isolate an infant for precautionary purposes and treatment.

The actual number of hospitals that have had the problem of staph infection in their nurseries as a result of "these directives" may never be known. Hospitals, with the exception of the Yale-New Haven Hospital in Connecticut, have been unwilling to report publicly an outbreak of staph in their nurseries and with good reason. Were a pregnant mother to read that staph had been reported in a given hospital, she might have serious misgivings about going to that hospital to have her baby.

In an article, "Staph Closes a Yale Nursery after Halt in Use of Germicide," the *New York Times* reported on February 1, 1972: "An outbreak of staphylococcal infection among newborn infants who had not been bathed with hexachlorophene has forced pediatricians to close one of three nurseries at the Yale-New Haven Hospital."

The only way the actual number of deaths from staph infection could be determined would be to analyze every death certificate throughout the nation. This would be an onerous if not impossible task. Staph infections in hospital nurseries may spread rapidly. Generally speaking, staph infections originate on the skin and are most commonly found in hospital nurseries.

By July 1972, sixty-two hospitals had quietly informed Sterling Drugs, Inc., the manufacturer of pHisoHex, of the presence of staph infection in their institutions after the discontinuance of the use of pHisoHex.

A news story headlined "Toiletry Firms Bow to Demand by FDA" appeared in the *San Jose Mercury* on January 7, 1972. The story stated: "FDA Commissioner Charles Edwards said that it [the FDA directive] was designed to protect the consumer from possible health hazards. FDA, he said, was unaware of 'any harm to humans from normal use of HCP' but was worried by studies showing that babies washed with HCP can absorb the chemical in their blood stream and that animals have developed brain lesions."

In an article, "Publishers Report, FDA Letters," in *Private Practice,* January, 1972, Dr. Francis M. Davis said, "It seems to me that the FDA too frequently acts as a result of pressures from 'consumer advocates' who are unfamiliar with the results of actual clinical use of a drug product. Relying on animal studies of questionable significance to the practicing situation, and a few speculative clinical articles, critical of either safety or efficacy, these propagandists with the cooperation of publicity-seeking Congressmen can apparently stampede the FDA into unwise action."

Robert M. Bleiberg, in a *Barron's* editorial (February 7, 1972), wrote the following: "After experiments with rats and monkeys in which the former were fed hexachlorophene and the latter immersed in it, FDA found traces of the substance in the blood and brain. These studies, stated the agency, 'challenged the safety of hexachlorophene bathing of infants,' a practice which it recommended be changed to soap and water. It also proposed that hexachlorophene products be put under prescription."

From the November 1972 *FDA Consumer* (the official magazine of the Food and Drug Administration): "The Food and Drug Administration restricted the use of hexachlorophene to products available on a doctor's prescription after scientific evidence indicated that the anti-bacterial chemical, when applied to the skin, can be absorbed into the bloodstream in toxic amounts."

The article further states, "The new evidence that confirmed the studies, and that led FDA in September 1972 to take even stricter regulatory actions than it had proposed nine months earlier, came essentially from two sources. Beginning in March 1972, babies in France were reported ill with an unusual combination of symptoms: rash, loss of appetite, sleepiness, and increasing irritability. The illness progressed to generalized muscular contractions and, for more than thirty babies, death.

"A careful study by French officials traced the illness to a batch of talcum power called Bébé, made by the long-established cosmetic manufacturer Morhange. Some samples of the powder contained 6 percent hexachlorophene.

"It is still not clear how the hexachlorophene got into the powder. One thing is certain: it was added inadvertently. The Bébé brand was not supposed to contain any hexachlorophene."

The tragic death of the French babies accounted for 50 percent of the FDA's reason for placing hexachlorophene on a prescription only basis. PHisoHex contains 3 percent hexachlorophene, *not* 6 percent, and it is rinsed off after usage. The talcum powder was not only applied at double strength (as compared to pHisoHex) but also it was not rinsed off and remained in contact with the baby's skin for an unknown period of time. The FDA has acknowledged the fact that the entire situation was accidental.

This situation is reminiscent of the tragic deaths that resulted from a bad batch of Salk polio vaccine by an American pharmaceutical company in the 1950s. It is fortunate that the FDA did not deprive millions of human beings of the benefits from the Salk vaccine on the basis of this tragedy. Instead the error was immediately traced and corrected.

Prior to the deaths of the French babies, the only evidence of toxicity to human beings from pHisoHex were such bizarre instances as pHisoHex having been drunk when it was accidentally mistaken for an antacid, when pHisoHex was drunk in place of a sulfa suspension, and when it was fatally applied to a severely burned person.

The FDA cited the following as its second reason for removing pHisoHex from the market: "Meanwhile, FDA received from the University of Washington at Seattle data which suggested that damage to the nervous system could be produced by 'routine' use of hexachlorophene—that is, from 3 percent solutions that were widely used in the United States. In the study, the brain tissues of babies who died from various natural causes were examined for microscopic abnormalities. Abnormalities appeared principally in the brains of babies who had been bathed in hexachlorophene, especially if they had been premature and had a number of baths.

"The conclusion reached by the University of Washington scientists was that there is a correlation between exposure of babies to hexachlorophene, and microscopic abnormalities in brain stems."

It is my opinion that the FDA's scant summary of the studies conducted at the University of Washington are misleading. The complete report by the three doctors on their facts, findings, and conclusions is thorough, informative, and revealing. This report has been submitted to the *Journal of the American Medical Association* for publication, and it is listed here under Alvord in the Special Reports Section of the List of References. With the permission of the Director of the project, the synopsis-abstract and the final concluding paragraph of their report are submitted in toto:

Exposure of infants to 3% hexachlorophene-bearing soap (pHiso-Hex[R]) is associated with a vacuolar encephalopathy of the brainstem tegmental reticular formation with a rostrocaudal distribution which varies with the age of the infant. The incidence of the lesion is related to the number of exposures and to the gestational age: lesions develop in 100% of infants weighing less than 1400 g [approximately 3 pounds] at birth and washed daily for 3 or more days with 3% hexachlorophene. In older infants the incidence is much less unless there is extensive exposure by a highly absorptive route such as abnormal skin. The elimination of hexachlorophene in very young premature infants (under 1400 grams birth weight) and in infants with skin defects and either limitation of the exposure of other infants to two baths or marked dilution of the soap may resolve the pediatricians' dilemma, eliminating most of the risk of neurotoxicity while preserving the antibacterial effectiveness in the newborn nursery.

Concluding paragraph:

Finally, we must face up to the fact that the clinical or functional significance of this lesion in these patients remains unknown. There were so many other complicating factors that the potential correlation of clinical worsening following pHisoHex[R] exposures has been difficult to establish to everyone's satisfaction. We are generally inclined to the views that abnormalities (lesions) are not good for individuals, that the medullary reticular formation is an especially important region of the brain for respiratory, vasomotor and oro-gastrointestinal control, and that the burden of proof must be on the individual who would suggest that these lesions are clinically insignificant. Until more is known about the antibacterial effectiveness of markedly diluted solutions of pHisoHex[R] and their neurotoxicity in small premature infants, we would suggest not using hexachlorophene in the small premature infant and markedly decreasing the amount used in less premature or full-term infants, probably by a hundred-fold dilution such as was routinely used at Children's Orthopedic Hospital [Seattle]. Strangely, however, we can find no published report suggesting the use of anything but undiluted pHisoHex[R] (10–12,16) [their bibliography], but the absence of serious infections at Children's Orthopedic Hospital suggests that marked dilutions can be effective as well as non-toxic.

The FDA and the news media placed heavy emphasis on unrealistic experimentation and the accidental death of thirty-five French babies while the crux of the entire situation lies in the case histories and autopsy reports prepared by neuropathologists at the University of Washington. Their findings proved beyond any reasonable doubt that hexachlorophene is dangerous for the bathing of *premature babies*. This meticulous work has been virtually ignored. These doctors suggested the continued use of hexachlorophene in diluted form on most full-term babies while they were *in the hospital nurseries*. This, too, has been ignored. Full-term babies are more susceptible to staph than premature babies because the premature babies are isolated, are frequently in incubators and are, of necessity, given more detailed care. Since staph is prevalent in *hospital nurseries* the "Washington team" has not recommended that hexachlorophene bathing be continued after the baby *leaves* the hospital.

Thousands of doctors and hospitals across the nation have discontinued the use of hexachlorophene, even by prescription, and the result at this writing is a six-fold or 500 percent increase of staph infection in hospital nurseries.

This has been brought about by accentuating the negative and eliminating the positive and is a tragic example of how the work of honest creditable scientists can and has been overshadowed by sensationalism.

The FDA was asked, "Is there any safe substitute for hexachlorophene?" Their answer was strangely familiar: "FDA's advisory committee on antimicrobial drugs is looking into the question, and its conclusions will be announced. No antibacterial will be permitted in any product unless its benefits outweigh its potential risks."

How many outbreaks of staph infection in hospital nurseries will there be in the interim? They speak of risks versus benefits, as indeed they should, but science without reason is science without meaning. In the case of hexachlorophene and the sudden rise in staph, it appears that the baby has quite literally been thrown out with the bath water.

Return
to Realism

12 By September, 1971, the case of DDT was "back on trial"—this time before a Federal Hearings Examiner, Edmund M. Sweeney, who heard testimony for seven months from 125 witnesses who submitted 300 documents. The entire proceedings resulted in some 8,900 pages of testimony and cost the taxpayer an estimated one-half million dollars.

Mr. Sweeney, an attorney for the United States Department of the Interior on special assignment to the EPA, reviewed all the testimony and submitted his recommendations of facts, findings, and proceedings to William D. Ruckelshaus, chief administrator of the EPA, on April 25, 1972. Basing his report upon the preponderance of reliable evidence, Mr. Sweeney recommended the continued use of DDT in most areas under consideration. On June 2, 1972, Mr. Ruckelshaus reversed the recommendations of the Hearing Examiner in virtually all areas.

Considering the length of the testimony and the length of both the Sweeney and Ruckelshaus reports, it is obviously impossible to review here every issue that was under consideration. It is, however, worth comparing the major issues as set forth in both the Sweeney and Ruckelshaus reports.

The basic issues in the 1972 EPA Hearings were defined by Mr. Sweeney: "(1) The nature and magnitude of the foreseeable hazards associated with DDT; and whether the hazard, if any, is inherent in the normal use of DDT or whether it results primarily from misuse; and (2) the nature of the benefit conferred by the use of DDT; whether its absence would merely cause some inconvenience to would-be users or would cause serious risks to public health or disruption of important social needs. Likewise [there must be considered] available alternatives and their propensities. . . . The Petitioners [for the continued use of DDT] have met fully their burden of proof."

From the Ruckelshaus Report: "The Pesticides Office of this Agency [Environmental Protection Agency] and intervenor Environmental Defense Fund (EDF) filed exceptions to the Examiner's report challenging his application of the burden of proof to this case, his findings of fact, conclusions of law, and numerous evidentiary rulings." This means that the EDF was allowed to file exception to a Federal Hearings Examiner's Report *prior* to an administrative decision which was to be based *upon* that report and before that report was released.

According to Examiner Sweeney, "The two most common allegations against DDT use are that it is detrimental to many non-target organisms, especially birds, fish, and crustaceans, and that it is possibly a carcinogen to man." Mr. Sweeney said that on the other hand the precipitous removal of DDT from interstate commerce could seriously disrupt public health programs and agricultural yields, and such action probably would result in the widespread use of highly toxic replacements. He said, "I believe that a balancing of benefit against risk if it has to err, should err on the side of safety; nevertheless, such a balancing must rest on solid grounds. The risk side should not be based on such factors as *possible* future risk, *unforeseeable* abuse, and a strict requirement of *no*-damage-to-wildlife. In my opinion, an *unreasonable* risk must be demonstrated in order to overcome the indicated or the proven benefits of DDT." (Emphasis added.)

The Hearings Examiner stated, "DDT is not a carcinogenic hazard to man. DDT is not a mutagenic or a teratogenic hazard to man. The uses of DDT under the registrations involved here do not have a deleterious effect on fresh water fish, estuarine organisms, wild birds, or other wildlife. The adverse effect on beneficial animals from the use of DDT under the registrations involved here, is not unreasonable on balance with its benefit." Mr. Sweeney further stated, "There is a present need for the continued use of DDT for the essential uses defined in this case."

The essential uses of DDT as defined in this case fall into two major categories: (1) crop uses—cotton, beans (dry, lima, snap), sweet potatoes, peanuts, cabbage, cauliflower, brussels sprouts, tomatoes, fresh market corn, sweet peppers, pimentos, onions, garlic, and commercial greenhouses; (2) non-crop uses—control of house mice and bats (military only), fabric treatment (military only), disease vectors, and quarantine control of body lice by prescription drugs.

The following is from Mr. Sweeney's summary: "Although it is not an issue here, there was ample evidence to indicate that DDT is not the sole offender in the family of pesticides; and that necessary replacements would in many cases have more deleterious effects than the harm allegedly caused by DDT.

"In my opinion, the evidence in this proceeding supports the conclusion that there is a present need for the essential uses of DDT; that efforts are being made to provide a satisfactory replacement for DDT; and that a co-operative program of surveillance and review can result in a continued lessening in the risks involved."

Mr. Ruckelshaus ignored Mr. Sweeney's conclusions and conversely stated: "I am persuaded for reasons set forth in Part III of this opinion [his written reversal of Sweeney's recommendations] that the long-range risks of continued use of DDT for use on cotton and most other crops is unacceptable and outweighs any benefits. Cancellation for all uses of DDT for crop production and non-health pruposes is hereby reaffirmed and will become effective December 31, 1972." Mr. Ruckelshaus did not cancel the use of DDT on onions, on green peppers, or on sweet potatoes in storage, or for use in public health.

The conflicts of opinions between the conclusions of these two men are difficult to comprehend and even more difficult to reconcile in light of the fact that they are based on the same testimony. In reference to the testimony on biological magnification (the build-up of a given material in living organisms), Mr. Sweeney stated, "The evidence indicates that there is no clear and concise understanding of this phenomenon and likewise the evidence does not clearly satisfy the curiosity as to whether such a food chain build-up actually has adverse effects (except in the situation where the concentration exceeds any promulgated tolerance level)." According to Mr. Ruckelshaus, "Persistence and biomagnification in the food chain are, of themselves, a cause for concern, given the unknown and possibly forever undeterminable long-range effects of DDT in man, and the environment."

Studies of the build-up of parathion and the closely related methyl parathion in the food chain show that these chemicals cannot biomagnify. The lethal effect of enough parathion is almost immediate, thus essentially terminating any build-up, except within the many scavenger organisms. Since sublethal amounts are quickly detoxified by susceptible organisms, surviving prey or predators will quickly dispose of ingested toxicants. Long-term feeding experiments using sublethal doses of parathion have shown that it does not cause tumors or affect the reproductive capacities of test animals. When a test animal is given enough parathion it dies long before it has time to develop a tumor let alone reproduce. This brings us to a ridiculous paradox: With a less toxic compound, larger, repeated doses can be administered in long-term experiments thus accentuating unfavorable results. Until this paradox is fully comprehended and realistic experimentation prevails, "What fools we mortals be."

On March 25, 1971, William D. Ruckelshaus testified at the Poage Hearings: "The protection of the environment is incredibly complex as a subject and it involves all kinds of considerations that are not usually taken into account when somebody talks about protecting the environment. The question of DDT is a very good example. To immediately suspend the use of DDT may cause more environmental problems than it solves immediately."

Yet on June 2, 1972, just fifteen months later, the same William D. Ruckelshaus suspended all uses of DDT except for three minor crops and public health use. He did comment within his 1972 report that his action was done with a scalpel instead of a hacksaw. (I would hate to see Mr. Ruckelshaus' hacksaw.) The Ruckelshaus Report was signed on June 2, 1972, but was not released to the press until June 14, conveniently after the last presidential primary for that year.

After considering a massive amount of evidence introduced by both sides, regarding the thin-eggshell controversy, Mr. Sweeney reported, "There was no evidence that DDT was the only factor in a decline of bird populations and examples given included pollutants, other pesticides than DDT, the advances of urbanization which brought human population changes and noise factors, and also disruption of the normal bird life by unnecessary intrusion."

Here again Mr. Ruckelshaus took issue with his Hearings Examiner's Report: "Laboratory evidence is also impressively abundant to show the acute and chronic effects of DDT on avian animal species and suggest that DDT impairs their reproductive capabilities. The Petitioners-registrants' assertion that there is no evidence of declining aquatic or avian populations, even if actually true, is an attempt at confession [confusion?]

and avoidance." Confession [confusion?] and avoidance of what?

Mr. Ruckelshaus said, "Group Petitioners also contend that the observed phenomenon of eggshell thinning and DDE residue data are tied by a statistical thread too slender to connect the two in any meaningful way. Viewing the evidence as a total picture, a preponderance supports the conclusion that DDE does cause eggshell thinning. Whether or not the laboratory data above would sustain this conclusion is beside the point." He might as well have said, "Don't confuse me with the facts, my mind is already made up."

From the Ruckelshaus Report: "The Agency and EDF [an Intervenor in the Hearings] have established that DDT is toxic to non-target insects and animals, persistent, mobile, and transferable and that it builds up in the food chain. No label directions for use can completely prevent these hazards. In short, they have established at the very least a risk of the unknown." Mr. Ruckelshaus went on to explain that these "unknown risks" to human health from using DDT cannot be discounted. Perhaps the fact that no human being has ever been harmed by DDT also falls into Mr. Ruckelshaus' category of being "beside the point." He virtually banned DDT because he believed it presented an unknown *risk* to human health. Yet he admitted that, in the event of an outbreak of malaria, yellow fever or typhus, DDT would be needed to protect human health.

By contrast Mr. Sweeney said, "Those that would ban all uses of DDT because of the possibility of some damage to man, the evidence of which is said to consist of the results of a few experiments with animals, would do well to compare such skimpy evidence of risk with the well-documented proof of the benefits which DDT has bestowed on mankind.

"Granted that the past successes of DDT in the area of public health cannot by itself justify any future use on agricultural crops, it must be noted that the DDT record of safety to man, particularly in comparison to other pesticides which might be used as replacements, is an item to be considered in any balancing of risks versus benefits in the case of DDT."

Before pursuing the conflicting reports of Mr. Sweeney and Mr. Ruckelshaus, it is worthwhile to pause briefly to examine what was *supposed* to have transpired. Obviously, Mr. Ruckelshaus, as chief administrator of the entire Environmental Protection Agency, could not possibly have spent seven months hearing testimony regarding just *one* pesticide out of the many because his duties encompass virtually all aspects of our environment. This is precisely why a Hearings Examiner (Mr. Sweeney) was appointed. Mr. Sweeney was to hear the entire testimony and then prepare a full written report based upon the facts brought forth in these hearings. This report was to serve as a factual guide from which Mr. Ruckelshaus would render a final decision. Established facts do not need support.

Mr. Sweeney carefully sifted out facts from fancy and presented them to the EPA chief, who ignored the facts and entertained the fancy. Mr. Ruckelshaus acted in a manner reminiscent of the Red Queen, whose solution to all problems was, "Off with their heads!"

Another important issue, that of insect specificity, was discussed by Mr. Sweeney, who said that no economic poison (and by definition DDT is an economic poison) is species specific. In other words, there is no pesticide that will kill only the target insect. Mr. Ruckelshaus: "There is no dispute on this record that DDT is a non-specific chemical that kills both target and non-target species in the immediate area of application. Few chemicals, however, are so selective that they can be used without causing some injury to 'non-target' species." Yet after making this statement, he listed the fact that DDT is lethal to many beneficial agricultural insects as one of his reasons for banning its use and advocated the use of methyl parathion as a substitute for DDT on most crops. Methyl parathion is immediately damaging to beneficial insects, not to mention birds, animals, farm workers, and children.

In reference to the effect of DDT on aquatic organisms Mr. Sweeney said, "DDT is not alone in detrimental effects on aquatic life. Other additions to the environment that affect marine fishes are heavy metals (zinc, cadmium), synthetic detergents, compounds with high biological oxygen demands, heated effluents from steam electric stations, effluent from pulp and paper mill waste, spills as a result of crude oil and their attempt to clean up mixtures with chemical oil dispersants." Mr. Sweeney continued, "Moreover, what problems may exist with respect to DDT's effect on aquatic life [are] due to abuse rather than use of DDT, as well as past accumulations from uses not presently at issue." In other words, we cannot go back to correct the errors of the past. During the years a great deal has been learned about the proper and safe use of pesticides. Mr. Sweeney concluded that although it is necessary to be vigilant, it is questionable whether the available evidence presents a present or foreseeable danger to our aquatic life from DDT.

The chief administrator of the EPA flatly stated, "DDT can have lethal and sublethal effects on useful freshwater aquatic invertebrates, including arthropods and molluscs. DDT is toxic to fish. DDT can affect the reproductive success of fish. DDT can have a variety of sublethal physiological and behavioral effects on fish."

The use of DDT on cotton was an important issue in the hearings. On the basis of the vast amount of evidence presented, Mr. Sweeney concluded, "While the evidence convinces me that the use of DDT on cotton is declining and should be reduced as soon as effective replacement means of controlling pests are developed, I do not feel that the evidence to date permits any conclusion to the effect that DDT should be banned for use on cotton at this time." Mr. Ruckelshaus banned the use of DDT on cotton.

The EPA Chief accepted methyl parathion as the substitute pesticide for DDT on most crops. His statements are baffling.

Early in the report he said, "The record before me leaves no doubt that the chief substitute for most uses of DDT, methyl parathion, is a highly toxic chemical and, if misused, is dangerous to applicators." He neglected to mention that it is also much more dangerous than DDT to field workers, pickers, children, and virtually all forms of wildlife. Mr. Ruckelshaus' ban allowed for a six-and-one-half-month period during which time he believed that farm workers could be educated to the safe use of this and other highly toxic pesticides. It would take a few *years* at best. As Dr. Gunther points out, "Even after twenty years of extensive use around the world, we still have deaths among applicators, pickers, and children from improper and incautious use of parathion and the related methyl parathion. Such deaths have occurred even among carefully trained workers." Meanwhile, entomologists are hard at work in test fields attempting to determine a safe time for re-entering those fields which have been sprayed with highly toxic insecticides. In light of the fact that a little child died from parathion poisoning after playing on a patio sprayed with parathion two years previously, the re-entry testing program, though meritorious, also could take years of extensive work. Furthermore, the re-entry program cannot prevent wildlife from entering a sprayed area before it is safe.

The purpose of this book is certainly not to dwell on trivialities; however, one admittedly trivial error by Mr. Ruckelshaus should give the reader cause to consider the need for a panel of qualified scientists, rather than political appointees with limited scientific backgrounds, to make environmental decisions. In reference to the re-education program, Mr. Ruckelshaus said, "Such a program can also introduce farmers to the less acutely toxic organophosphates, like carbaryl, which may be satisfactory for many uses." Carbaryl is *not* an organophosphate. It is a carbamate, with the trade name SEVIN, and it is not even remotely related to the organophosphates. If a journalist knows this distinction, certainly the protector of our environment should be aware of it.

Mr. Ruckelshaus made some strange statements regarding methyl parathion. He said, "Methyl parathion and other organophosphate chemicals are effective for the control of cotton pests. Methyl parathion and organophosphates are less toxic to aquatic life than DDT. [Methyl parathion *is* an organophosphate.] Methyl parathion and organophosphates appear to be less 'persistent' and do not build up in the food chain." Further into the report Mr. Ruckelshaus said, "Many poisonings have been attributed to the use of methyl parathion. Untrained users of methyl parathion are frequently not sufficiently careful in its use despite label directions." Many itinerant farm workers cannot read and if they can read the terminology is usually incomprehensible to them. Also, many farm workers are Mexican-American and can read only Spanish. Animals and little children cannot read any language. Mr. Ruckelshaus continued by stating, "Methyl parathion can be used safely." Yet, on the same page, he said, "Methyl parathion is dangerous to users and presents a risk to them."

Methyl parathion has an acute dermal toxicity level in both male and female rats of 67 milligrams per kilogram (mg/kg) of body weight, while DDT has a corresponding level of only 2510 mg/kg. The acute oral toxicity level of methyl parathion for rats is 14 mg/kg in males and 24 in females, while that of DDT is 113 mg/kg in males and 118 in females. In spite of these facts, Mr. Ruckelshaus accepted methyl parathion as a replacement for most uses of DDT.

The subject of c-ncer in relation to DDT was an emotion-packed part of the Hearings. Having heard the entire testimony, Mr. Sweeney concluded, "DDT is not a carcinogenic hazard to man." He said that he had given considerable weight to the testimony of the Surgeon General which Mr. Ruckelshaus termed "cautious." Within his testimony, Surgeon General Jesse Steinfeld stated, "We have no information on which to indict DDT either as a tumorigen or as a carcinogen for man and on the basis now available, I cannot conclude DDT represents an imminent health hazard." Prior to his appointment as

Surgeon General, Dr. Steinfeld did extensive work in the field of cancer research. Since Mr. Ruckelshaus questioned and discounted the Surgeon General's testimony, Mr. Ruckelshaus' statements on the carcinogenic effects of DDT on human beings bear examining. Early in his report he stated, "The possibility that DDT is a carcinogen is at present remote and unquantifiable; but if it is not a siren to panic, it is a semaphore which suggests that an identifiable public benefit is required to justify continued use of DDT." Later in the report he said, "DDT is a potential human carcinogen." *On the same page* he said, "There is no adequate human epidemiological data on the carcinogenicity of DDT, nor is it likely that it can be obtained." *Four lines later* attorney Ruckelshaus, who had the audacity to question the expertise of a former cancer researcher and the present Surgeon General of the United States, stated unequivocally, "DDT presents a carcinogenic risk."

It is difficult, to say the least, to understand the basis for the Ruckelshaus decision. Basically the Hearings were to determine whether DDT presented either a public health hazard or an undue risk to our environment. Mr. Ruckelshaus has acknowledged, in reference to the food chain, that the risk is unknown. He carefully avoided any mention of DDT in mother's milk. He leaned heavily upon DDT's alleged toxicity to fish, ignoring the fact that in the few instances where DDT did result in fish kills, it was a result of misuse or accidental overuse of DDT—a problem that has since been corrected. He persisted in claiming that DDT affected the reproductive success of fish in spite of the fact that testimony to this effect supplied by EDF's Dr. Philip Butler was proved to be greatly distorted. He ignored the fact that the Surgeon General and many other cancer specialists firmly stated that DDT was not a carcinogen to man. In fact, Mr. Ruckelshaus admitted that the possibility that DDT is a carcinogen is remote and unquantifiable. Yet at the same time, he declared DDT to be a carcinogenic risk to man.

In the final analysis, he accepted the use of methyl parathion for most uses of DDT, in spite of the fact that he was admittedly aware of the risk it presents to human beings and was apparently oblivious to the fact that it also presents a serious threat to all forms of wildlife in areas sprayed by this pesticide.

For the above-mentioned reasons Mr. Ruckelshaus, in the name of environmental protection and public safety, reversed Mr. Sweeney's recommendations—unless of course there should be an outbreak of malaria, typhus, or some other insect-borne disease. Should this happen, our environmental protector reserved the right to protect the public health with the material he had already declared to be a menace.

Within minutes after the Ruckelshaus decision was released, an attorney for the twenty-seven petitioners for the continued use of DDT filed an appeal in the Fifth Circuit Court of Appeals in New Orleans. Almost simultaneously the EDF filed an appeal in the U.S. Court of Appeals in Washington, D.C. This EDF appeal sought to have the use of DDT discontinued on even the three minor crops. Considering the heavy court dockets and the maze of our judicial system, the DDT controversy could drag on for years and may ultimately be decided by the Supreme Court.

Should the Supreme Court uphold a ban on the production, use, and interstate shipment of DDT, such a decision would be binding on all states. Should the Supreme Court decide in favor of the continued production and use of DDT, that would mean only that it *can* be used, not that it *must* be used. The decision to use or not to use DDT would then revert to the separate states, counties, and even cities.

While the nation and the world awaited a decision at the Federal level, voters in the state of California were bombarded by advertising campaigns designed to woo their vote for Proposition 9, which appeared on the California Primary ballot in June, 1972. There were twenty-three sections of legislation tucked neatly into this one grossly misleading initiative. Proposition 9 was referred to in California as the "Clean Environment Bill." The following words appeared on the California ballot: "Provides specific air pollution standards for autos and businesses. Restricts oil and gas extraction, use of pesticides, and construction of nuclear power plants. Imposes civil and criminal penalties for violations."

Yet the actual wording (regarding pesticides) as stated in the body of the bill read, "Prohibits any person from manufacturing, possessing, buying, selling, importing, delivering, or using any form of persistent chlorinated hydrocarbons, as defined within this state, unless pursuant to a permit issued by the Director of Agriculture. Provides that such permit may be issued only upon authorization by four-fifths vote of the members of each house of the Legislature."

Rarely is there four-fifths attendance in either the California Assembly or the California Senate even when their salaries are on the docket. The proposition was soundly defeated.

Since Montrose Chemical Company is located in California and is the *only* company in the United States that manufactures DDT, and since approximately 97 percent of this DDT is exported for use in the developing nations in the fields of agriculture and public health, this initiative, had it passed, would have become an international matter.

Mr. Ruckelshaus' decision concerned only one pesticide, DDT, while the California initiative was directed at *all* chlorinated hydrocarbon pesticides.

It is worthwhile, particularly in light of the foregoing material, to restate a portion of Dr. Norman Borlaug's testimony at the EPA Hearings: "The indiscriminate cancellation, suspension or outright banning of such pesticides as DDT, is a game of dominoes we will live to regret.

"DDT, because it is a name popularly known to most segments of the public, has been the first target. Once this is accomplished, the so-called ecologists will work on hydrocarbons, then organophosphates, carbamates, weed-killers, and, perhaps, even fertilizers will come under the assault of their barrage of misinformation."

At this writing the manufacturers, formulators, and agricultural users of toxaphene have been required to submit written evidence to the EPA to justify its continued use. According to Dr. Gunther: "Toxaphene belongs to the same chlorinated hydrocarbon (organochlorine) group as DDT. It is another broad-spectrum insecticide especially useful for the control of insect pests on cotton, on many field crops, and on domestic animals. Two other similar hard organochlorine insecticides under fire as 'undesirables' are chlordane (cockroaches, ants, termites, soil insects, and some insect pests of field and vegetable crops) and lindane (grasshoppers, seed treatment, soil insects, some cotton insect pests). A separate book could be written on the necessity of having some 'hard' pesticides available. They are 'hard' because they *do* persist in the environment for longer periods. Imagine treating a house for termites every two weeks, or spraying mud huts weekly for mosquito control, over a period of years with the highly toxic organophosphates which require trained applicators wearing protective clothing and face masks."

In the October 5, 1969, issue of the *Seattle Times,* Dr. Wurster was quoted as follows: "If the environmentalists win on DDT, they will achieve, and probably retain in other environmental issues, a level of authority they have never had before. In a sense, then, much more is at stake than DDT."

In the *Reader's Digest,* October, 1969, EDF's same Dr. Wurster stated: "Actually, I think the reason the industry is fighting so hard to save DDT is not because it's better or cheaper, but because its banning could establish what would be from their point of view a dangerous precedent. Up to now, the regulation of pesticides has been almost solely in the hands of agriculture-oriented interests—chemical companies, farmers, departments of agriculture. They're afraid to let environmental scientists get a foot in the door, because we're only secondarily interested in the economics of farming."

One need have no connection or vested interest in chemical companies, agriculture, or politics to fear letting these people "get their foot in the door"—or letting them decide what is good for agriculture, for this country, or for the world.

Money is desperately needed for research and development in the field of agricultural chemicals. An article titled "Pesticide Economics Moving in New Directions," in the April, 1972, issue of *Chemical and Engineering News,* discussed the problems of those companies which manufacture agricultural chemicals. According to this article, industry has begun to reduce its expenditure for research and development in the field of agricultural chemicals, and for good reason. Out of every 10,000 compounds tested, on an average only one is found to be successful. Research and development costs for a new pesticide, from test tube to pilot plant production, are estimated at ten million dollars. The length of time between the initial testing and the marketing is usually from eight to ten years.

In order to be profitable to the company, "the winner" pesticide must last at least nine years at an annual sale level of ten to twenty million dollars per year, with a return on investment of 40 percent before taxes. Considering public pressure and insect resistance, few pesticides meet these standards. Obviously, were a given company to invest the originally cited ten million dollars at 8 percent interest, it not only would have the assurance of a secure investment, but also would make more money over the ten-year period from interest on the principal.

According to Dr. Gunther, "In the United States, twenty years ago, there were at least fifty chemical companies working to develop new pesticides. Now there are only ten or twelve because the risk is so high—high cost of research and development, low probability of success and low return on the investment. Each year the boards of directors of the few remaining companies ask, 'Shouldn't we get out of the pesticide field?' "

No pesticide is effective against all insects. There is no pesticide with which insect resistance has not been a problem. DDT is certainly no exception, though it does have an amazing record of use and safety over a thirty-year period.

DDT is now useless against the common housefly and many other household pests. There is no possibility that this pesticide could be used indefinitely. Considering the time and the amount of money required to develop new pesticides, it is vital that research and development be accelerated.

A significant example of insect resistance to pesticides was seen in the tobacco fields in the lower Rio Grande Valley in Texas. In 1968, the tobacco budworm began to show early signs of resistance to organophosphate pesticides. Many fields were treated fifteen to eighteen times with methyl parathion, and even then the growers suffered serious losses. According to Dr. Perry L. Adkisson, entomologist at Texas A & M, "By 1970 the insect could not be controlled regardless of the insecticide used, and the valley that year produced the lowest yield recorded the past twenty-five years."

On March 12, 1972, the *New York Times* printed an article by the critic Brooks Atkinson under the headline, "A Part of America That is Dying": "No change in the condition of the bald eagle. The bald eagle population which consists of about 750 breeding pairs in the contiguous United States continues to decline. Alexander Sprunt IV, Director of the National Audubon Society Research Program, believes that the population has declined about 10 or 15 percent during the last decade." Considering that 267 bald eagles were counted on *one* day on *one* creek in *one* state (Montana) in 1971, the decline mentioned is highly questionable.

Mr. Atkinson went on to say, "There are few in such favored habitats as Maine, the Chesapeake Bay area, and the Great

Lakes. In some places the bald eagle has vanished. Before World War II and the proliferation of insecticides, the bald eagle used to fish off the ice every winter in the Hudson River north of the George Washington Bridge. No more. Mr. Sprunt dogmatically declared, 'Unless we ban DDT the bald eagle will become obsolete.' The emblem on the Great Seal of the United States represents a part of America that is dying."

Mr. Atkinson refers to areas where the devastation of the gypsy moth, as a result of the *lack* of DDT, has had catastrophic effects on *all* forms of wildlife. In regard to the area north of the George Washington Bridge in New York, the human population on the New York side of the Hudson River is eight million people. On the New Jersey side of the river is the Palisades Parkway, one of the busiest freeways in the area and built, incidentally, since World War II. The Palisades Parkway runs through the Palisades Interstate Park, a strip of land only a few hundred yards in width that extends intermittently for 40 miles along the river. The park teems with cars and people. Naturally the bald eagle has moved from that area, but this certainly does not mean that a part of *America* is dying and the implication should be resented.

In the next paragraph Mr. Atkinson wrote: "Since the eagle has for centuries symbolized power and majesty, there is something ironic about debasing it into an endangered species now. Insecticide poisoning seems like a squalid way of destroying a royal bird. Eagles appeared on Greek and Roman coins and on the battle standards of Roman and Napoleonic Armies. The power of the eagle's attack, the ferocity of his beak, his aloof and proud posture and the majesty of his flight appealed to the psyche of ambitious nations that would like to be imperialistic without using that distasteful word. The life style of the bald eagle is imperialistic."

Here, again, is an example of a writer using emotionalism rather than straightforward scientific evidence in an attempt to influence his reader.

The time is long overdue for placing agricultural chemicals in their proper perspective. The traditional cliché "Better late than never" applies. The use of chemicals has somehow become divided into two categories—the good and the bad. The basic problem seems to be our inability to avoid reacting by extremes, brought about by pressure and emotionalism.

In the minds of many people chemicals, in the form of pharmaceutical products, such as aspirin, penicillin, and insulin, are good; yet, pesticides and fertilizers, which *prevent* disease and increase our food supply, are frowned on. It has been interesting to note, in the course my two and a half years of research on DDT, just how many people consume aspirin, tranquilizers, antibiotics, and thousands of other drugs, while at the same time condemning the use of chemicals.

It is vital to the well-being of all mankind that we return to the middle of the road and take a realistic, rather than an emotional, stand on the use of pesticides and fertilizers. By so doing, hopefully the day will come when people, regardless of where they live, will have the benefits of a sufficient and well-balanced diet plus adequate protection against insect-borne disease. This can only be achieved through the application of chemistry—in perspective.

Recommendations

1. That Section 402 of the Food, Drug, and Cosmetic Act and Section 409 (The Delaney Clause) of the Food Additives Amendment be replaced by new legislation that would realistically protect the public as it was originally intended to do, while at the same time protecting that same public from unnecessary or harmful restrictions brought about by unscrupulous and misleading experimentation.

2. That the FDA require, when a warning label is deemed necessary, that the wording of such a label be written so that it can be readily understood by the user, and that, in addition, it be written in Spanish as well as English, since Spanish is the language spoken by the majority of non-English-speaking Americans.

3. That extensive Federal funds be made available to the land-grant colleges and universities for large-scale and continued experimentation and testing of agricultural pesticides and fertilizers. Such facilities and personnel exist now but are *seriously* hampered by lack of money. An additional advantage to more extensive utilization and substantial Federal financing of these facilities is that they are present in every state and

therefore are better able to specialize in problems indigenous to their particular areas.

4. That appropriate tests be required to show beyond a reasonable doubt whether a chemical is harmful *or* beneficial to man and/or the overall environment. Food additives, pesticides and fertilizers should be judged by a panel of experts to determine their usefulness and safety, and not be convicted or acquitted by those not qualified in the particular area of study.

5. That both State and Federal governments establish agricultural review boards composed of qualified scientists to advise them on the use of agricultural chemicals and proposed legislation regarding such chemicals in order that they may establish a reasonable balance of risk versus benefits. Further that such review boards advise regarding the use of such chemicals in the home, in the garden, in structural pest control, and in the commercial production of ornamental plants.

6. That prior to banning any chemical that is generally beneficial and not shown to be detrimental except in rare and isolated instances, a suitable replacement be readily available to people of all nations, in adequate quantity, and at a comparable price.

7. That such organizations as the EDF be required to show new evidence that will withstand scientific examination before additional tax money, that could be used for research, is spent on repeated hearings for DDT or any chemical.

8. That no ban on the use, production, or interstate shipment of DDT be allowed, but rather that the uses of this pesticide and all pesticides be restricted to specific uses in designated amounts, as determined by a panel of experts.

9. That we work toward achieving a properly paced "imbalance" of nature in man's favor, with our primary goals being the protection and prolongation of human life, not only in this country but throughout the world. At the same time, we must make every effort to protect our wildlife and our overall environment.

10.   That those citizens who are vitally concerned with their environment as well as with public health and safety, write to their congressmen at both the State and Federal levels to urge them to intercede in behalf of reasonable rather than emotional legislation regarding agricultural chemicals.

Selected List of References

I have chosen to document all sources in the text of this book rather than to encumber the reader with footnotes. For those who wish more detailed information I have prepared the following list of books, journal articles, government documents, and unpublished reports, some of which have been used in the text and a few of which have served as background information only. These references contain a cross-section of the pro and con arguments about DDT and other topics discussed in *The DDT Myth*.

BOOKS

Carson, Rachel, *Silent Spring,* Houghton Mifflin Company, Boston, 1962.

Duffy, J., *Epidemics in Colonial America,* Louisiana State University Press, Baton Rouge, 1953.

Ehrlich, Paul R., *The Population Bomb,* Ballantine Books, New York, 1968.

Godfrey, Arthur, *The Arthur Godfrey Environmental Reader,* Ballantine Books, New York, 1970.

Graham, Frank, Jr., *Since Silent Spring,* Fawcett Publications Inc., Greenwich, Conn., 1970.

Gunther, Francis A., and Blinn, Roger C., "Persisting Pesticide Residues in Plant Materials," *Annual Review of Entomology,* Annual Review, Inc., Stanford, Calif., 1956.

Gunther, Francis A., and Jeppson, L. R., *Modern Insecticides and World Food Production,* Chapman and Hall, London, 1960.

Manson-Bahr, P. H., *Manson's Tropical Diseases,* 13th edition, Cassell and Company, London, 1950.

Still, Joseph W., *Science and Education at the Crossroads,* Public Affairs Press, Washington, D.C., 1962.

Warshaw, L. J., *Malaria: The Biography of a Killer,* Rinehart and Company, Inc., New York, 1949.

Whitten, Jamie L., *That We May Live,* D. Van Nostrand Reinhold Company, Princeton, N.J., 1966.

JOURNAL ARTICLES

Arena, J. M., "Contamination of the Ideal Food," *Nutrition Today,* Vol. 5, No. 4, Winter 1970, pp. 2–8.

Brindley, William A., "Some Considerations—Pesticides as Chemicals," *Utah Science,* Agricultural Experiment Station, Utah State University, Logan, June 1971, pp. 59–61.

Flieger, B. W., "A Fish Story—Anadromous, i.e.," *Pulp and Paper Magazine of Canada,* September 20, 1968, pp. 3–8.

Graham, J. E., "Some Benefits of Pesticides to Public Health," *Utah Science,* Agricultural Experiment Station, Utah State University, Logan, June 1971, pp. 50–52.

Heath, Robert G., Krutzer, J. F., and Span, James W., "Marked DDT Impairment of Mallard Reproduction in Controlled Studies," *Nature,* Oct. 4, 1969, pp. 47–48.

Hertz, Karl O., "The Case of MSG—A Most Significant Guinea Pig," Editorial in *Food Technology,* Jan. 1970, pp. 13–14.

Laws, E. R., Jr., "Evidence of Antitumorigenic Effects of DDT," *Archives of Environmental Health,* Fall 1971.

Laws, E. R., Jr., "Man with Extensive Occupational Exposure to DDT," *Archives of Environmental Health,* Vol. 15, Dec. 1967.

Leposky, George, "The Alewives Invasion," *Animal Kingdom,* Dec. 1969, pp. 25–29.

McNulty, Faith, "The Silent Shore," *Audubon Magazine,* Nov. 1971, pp. 5–11.

Neuhold, John M., "Pesticides and Ecology," *Utah Science,* Agricultural Experiment Station, Utah State University, Logan, June 1971, pp. 39–41.

Olsen, Lyman J., "Pesticides in Public Health," *Utah Science,* Agricultural Experiment Station, Utah State University, Logan, June 1971, pp. 45–46.

Ortelee, M. F., "Study of Men with Prolonged Intensive Occupational Exposure to DDT," *Archives of Industrial Health,* Vol. 18, pp. 433–440, 1958.

Roberts, Reed S., "A Brief History of Pesticide Use," *Utah Science,* Agricultural Experiment Station, Utah State University, Logan, June 1971, pp. 42–44.

Tarrant, K. R., and Tatton, J. O. G., "Organochlorine Pesticides in Rainwater in the British Isles," *Nature,* Vol. 29, Aug. 17, 1968, pp. 725–726.

Tucker, R. K., "Chlorinated Hydrocarbons Cause Thin Egg Shells but So May Other Pollutants," *Utah Science,* Agricultural Experiment Station, Utah State University, Logan, June 1971, pp. 47–49.

Utah Science, editorial, "Some Considerations—Our Jekyll and Hyde Chemical World," Agricultural Experiment Station, Utah State University, Logan, June 1971, pp. 35–38.

Woodwell, G. M., "Toxic Substances and Ecological Cycles," *Scientific American,* March 1967, p. 9.

Wurster, Charles F., "DDT Reduces Photosynthesis by Marine Phytoplankton," *Science,* March 29, 1968, pp. 1474–1475.

GOVERNMENT DOCUMENTS AND RELATED REPORTS

Edwards, J. Gordon, Written Statement to California Water Quality Control Board, Sacramento, California, March 6, 1971.

Edwards, J. Gordon, Written Statement to DDT Hearings Officer, Washington State Department of Agriculture, Seattle, Washington, Oct. 16, 1969.

Elson, P. F., "Effect on Wild, Young Salmon of Spraying DDT over New Brunswick Forest," *Journal Fisheries Research Board of Canada,* Vol. 24, No. 4, 1967, pp. 731–761.

Gress, Franklin, "Reproductive Status of the California Brown Pelican in 1970 with Notes on Breeding Biology and Natural History," *Wildlife Management Branch Administrative Report* No. 70–6, July 1970. Supported by Federal Aid in Wildlife Restoration Project W-54-R, "Special Wildlife Investigation."

Hayes, Wayland J., Jr., "Monitoring Food and People for Pesticide Content," *Scientific Aspects of Pest Control,* Publication 1402, National Academy of Science, National Research Council, Washington, D.C., 1966.

Los Angeles County, Sanitation Districts, "Pesticides and Heavy Metals," *Progress Report,* Dec. 1970.

Paarlberg, Don, "Norman Borlaug—Hunger Fighter," U.S. Government Printing Office, No. O - 412–097, Dec. 1970, 20 pp.

Ruckelshaus, William D., Report on EPA Consolidated DDT Hearing, Environmental Protection Agency, I. F. & R. Docket NOS. 63 et al., June 2, 1972. (Copy courtesy of Montrose Chemical Company.)

Sobelman, Max, Editor, "Selected Statements from State of Washington DDT Hearings" (Seattle, Oct. 14, 15, 16, 1969), Publication of Montrose Chemical Company, Library of Congress Catalog No. 77–12457, MARC Classification Number QH 545.P456, Dec. 1970.

Sweeney, Edmund M., Hearing Examiner of EPA Consolidated DDT Hearing, *Hearing Examiner's Recommended Findings, Conclusions and Orders* (40 CFR 164.32), issued April 25, 1972. (Copy courtesy of Montrose Chemical Company.)

U.S., Congress, House, Committee on Agriculture, *Federal Pesticide Control Act of 1971,* Hearings (Chairman, W.R. Poage), 92d Cong., 1st Sess., U.S. Government Printing Office, Washington, D.C., No. 58–542, 1972.

U.S., Congress, House, Committee on Appropriations, *Agriculture—Environmental and Consumer Protection Appropriations for 1973, Hearings* before a subcommittee (Chairman, Jamie L. Whitten), 92d Cong., 2d Sess., U.S. Government Printing Office, Washington, D.C., No. 78–655 O, 1972.

U.S., Congress, Senate, Committee on Agriculture and Forestry, *Federal Environmental Pesticide Control Act, Hearings* before a subcommittee (Chairman, Herman E. Talmadge), 92d Cong., 1st Sess., U.S. Government Printing Office, Washington, D.C., No. 59–044 O, 1971.

U.S., Department of Agriculture, Forest Service, *Draft Environmental Impact Statement on the 1972 Cooperative Gypsy Moth Suppression and Regulatory Programs,* Presented to Russell E. Train, Chairman, Council on Environmental Quality, Dec. 13, 1971.

U.S., Department of Commerce, National Oceanic and Atmospheric Administration, National Marine Fisheries Service, *Fisheries of the United States 1970,* Current Fishery Statistics No. 5600, U.S. Government Printing Office, Washington, D.C., March 1971.

U.S., Department of Health, Education, and Welfare, *Report of the Secretary's Commission on Pesticides and Their Relationship to Environmental Health, Parts I and II* (Chairman, Emil M. Mrak), Dec. 1969.

U.S., Environmental Protection Agency, *Consolidated DDT Hearing,* Washington, D.C.—Arlington, Va., Fall 1971-Spring 1972 (Hearing Examiner, Edmund M. Sweeney). (Transcript of Hearing courtesy of Montrose Chemical Company.)

SPECIAL REPORTS

Alvord, Ellsworth C., Jr., Leech, Richard W., Shuman, Robert M., "Hexachlorophene Neurotoxicity" (Running title of paper, submitted for publication to *The Journal of the American Medical Association*), University of Washington School of Medicine, Seattle, Washington.

Borlaug, Norman E., "The Green Revolution, Peace and Humanity," Lecture on the Occasion of the Award of the Nobel Prize, 1970, Oslo, Norway, December 11, 1970. Original Copy entitled, "Reimpression De LES PRIX NOBEL EN 1970." Copyrighted by the Nobel Foundation, 1971. (English copy courtesy of Dr. Borlaug.)

Borlaug, Norman E., "Mankind and Civilization at Another Crossroad." Speech delivered on October 26, 1971. (Copy courtesy of Dr. Borlaug.)

Furrer, John D., "Pesticide Use and the Impact of Their Elimination," unpublished. Available USDA Extension Service, University of Nebraska, Lincoln.

Sobelman, Max, editor, "Selected Statements from State of Washington DDT Hearings" (Seattle, Oct. 14, 15, 16, 1969), Publication of Montrose Chemical Company, Library of Congress Catalog No. 77–12457, MARC Classification Number QH 545.P456, Dec. 1970.

Index